Essentials of Positioning and Location Technology

Mystified by locating and positioning technologies? Need to get the best from your location system? This guide is invaluable for understanding how the positions and movements of objects can be measured and used for real-world applications. From it, you'll learn how to optimise and manage system performance by working with parameters such as velocity, orientation, time, proximity and direction, and consider not only accuracy, but also reliability, integrity, response time and uncertainty.

Packed with practical examples, this concise book gives you an overview of terrestrial radiolocation techniques, including comparative system architectures and real-world performance and limitations. It describes inertial navigation principles and techniques, including low-cost MEMS sensors for consumer products, and a range of applications, such as those benefiting from hybrid positioning techniques.

DAVID BARTLETT is a chartered electronics engineer and serial entrepreneur, and is currently CTO and Director of Omnisense, a Cambridge-based company, which he co-founded. He has founded or co-founded several companies, is named inventor on many patent families, and is author of a number of published articles.

The Cambridge Wireless Essentials Series

Series Editors
WILLIAM WEBB *Neul, UK*
SUDHIR DIXIT *HP Labs, India*

A series of concise, practical guides for wireless industry professionals.

Martin Cave, Chris Doyle and William Webb, *Essentials of Modern Spectrum Management*

Christopher Haslett, *Essentials of Radio Wave Propagation*
Stephen Wood and Roberto Aiello, *Essentials of UWB*
Christopher Cox, *Essentials of UMTS*
Steve Methley, *Essentials of Wireless Mesh Networking*
Linda Doyle, *Essentials of Cognitive Radio*
Nick Hunn, *Essentials of Short-Range Wireless*
Amitava Ghosh and Rapeepat Ratasuk, *Essentials of LTE and LTE-A*
Abhi Naha and Peter Whale, *Essentials of Mobile Handset Design*
David Bartlett, *Essentials of Positioning and Location Technology*

For further information on any of these titles, the series itself and ordering information see www.cambridge.org/wirelessessentials

Essentials of Positioning and Location Technology

David Bartlett
Omnisense, Cambridge

CAMBRIDGE UNIVERSITY PRESS
Cambridge, New York, Melbourne, Madrid, Cape Town, Singapore, São Paulo, Delhi, Mexico City

Cambridge University Press
The Edinburgh Building, Cambridge CB2 8RU, UK

Published in the United States of America by Cambridge University Press, New York

www.cambridge.org
Information on this title: www.cambridge.org/9781107006218

First published 2013

Printed and bound in the United Kingdom by the MPG Books Group

A catalogue record for this publication is available from the British Library

Library of Congress Cataloguing in Publication data
Bartlett, David, 1958– author.
Essentials of positioning and location technology / David Bartlett, Omnisense, Cambridge.
  pages   cm. – (The Cambridge wireless essentials series)
ISBN 978-1-107-00621-8 (hardback)
1. Location-based services.   2. Mobile geographic information systems.   I. Title.
TK5105.65.B37   2013
910.285–dc23
                                                                    2012035240
ISBN 978-1-107-00621-8 Hardback

Contents

'This is a comprehensive and approachable introduction to positioning systems that draws together in one book the wide variety in the subject. David Bartlett's no-nonsense style gives us the essence of complex and sometimes abstract ideas in an easily digestible form, backed up with plenty of references where the reader can find more information. I very much enjoyed reading the book, and think that experts and beginners alike can find here much to engage and inform.'

Peter Duffett-Smith, Cavendish Laboratory, University of Cambridge

1 Introduction

Location is becoming an important and integral part of our everyday lives, spurred on by the widespread, almost ubiquitous, availability of GPS (Global Positioning by Satellite) technology in everyday consumer devices such as car navigation systems, mobile phones and cameras, and the increasing adoption of inertial sensors, particularly accelerometers and magnetometers in everyday products. In order to set the context, this book first looks at coordinate systems and what is meant by position or location and how to describe this information [Chapter 2].

Because of the importance of GPS, the next topic covered is global navigation satellite systems (GNSS) [Chapter 3]. The ability to determine the precise location of a device anywhere in the world is in turn leading to the emergence of many different Location-Based Services promoted by leading global companies such as Google with Google Maps and Latitude, Nokia, Microsoft, AOL and community initiatives such as OpenStreetMap.

However, despite the hype surrounding GPS it is not the only positioning technology available and indeed there are many applications for which it doesn't offer the required capability or performance. This book is intended to give a clear understanding of the different options available for locating and positioning systems with an emphasis on their real-world capabilities and applications. The next chapter [Chapter 4] covers the most important methods for determining position using radio signals, then Chapter 5 covers inertial navigation techniques and Chapter 6 looks at other methods of locating and positioning things. Chapter 7 deals with accuracy and performance and what they mean, as well as some of the fundamental techniques relating to location and position. Of course things never go entirely as planned, so Chapter 8 considers errors and failures and how to deal with them.

Having looked at the different technologies, architectures, systems and solutions available the book rounds off this essential foundation with a discussion of selected applications [Chapter 9], followed by a quick look at emerging opportunities and future challenges [Chapter 10].

1.1 A brief historical perspective

This book is intended to provide the essential things you need to know about locating and positioning technology and applications today. This chapter includes an overview of the different methods, techniques and technologies used for determining location, but it is not intended as a treatise of the history of navigation. There are several good books that describe the history of navigation and, despite being primarily a broad text about Global Positioning Systems, Samama [1] gives an excellent historical overview that could be a good place to start for many readers interested in the historical aspects of modern positioning technology.

1.2 What is meant by location or positioning?

The terms Location and Position are sometimes used interchangeably. For example we refer to the Global Positioning System (GPS), but we talk about Location-Based Services (LBS). The locations used in the latter are very often GPS positions.

However, location and position require more than just an X, Y, Z point in space or a latitude and longitude: time, velocity, orientation and other parameters describing the point may also be relevant. In this book we will sometimes use the terms Location and Position interchangeably; this simply reflects that they are to a large degree synonyms. Being pedantic we interpret Location as usually referring to a geographic position, often best described as a latitude, longitude and optionally height. On the other hand a position is more generic and may describe relative position and orientation of objects, not necessarily using the global coordinate system of latitude and longitude, and usually including time and often direction information.

The basic measure of a position or location is its place in space, typically described using Cartesian coordinates (X, Y, Z), or latitude, longitude and altitude. A detailed look at coordinate systems is presented in Chapter 2.

Since positions are transient, the basic position information is usually associated with a time at which it is valid.

However, positioning an object is often only complete if we also know its orientation – for example the direction one is facing.

Therefore to fully describe the position of an object in 3D one ideally needs at least seven parameters: three position, three orientation and time.

It is also often advantageous to include some of the derivatives of the basic positional measures, in particular velocity. Velocity can be described as a Cartesian vector or alternatively as a speed and direction. In either case three parameters are needed to fully describe velocity in 3D. Sometimes the derivatives of orientation being rate of rotation can be useful and similarly the second derivatives of position – acceleration. These are particularly useful when looking at inertial navigation techniques and for dynamic modelling of an object's motion. More detailed descriptions of inertial techniques will be covered in Chapter 5.

1.3 Describing a position

There are many different ways of describing the position of an object, depending on how the information is to be used.

In a navigation context the location of an object is usually described using latitude, longitude and height based on a suitable coordinate system. This method was invented by mariners for navigation at sea in order to know where they were located in a global context. It is the format generally used by GPS receivers today – they usually provide an output in latitude, longitude and height which defines a unique location relative to the planet as whole, although this is not the coordinate system used for carrying out the GPS position calculations. Chapter 2 provides an in-depth description of global coordinate systems and some of the intricacies involved in using them.

For our day-to-day activities a latitude and longitude is often not very helpful. As humans we usually use position information in a contextual manner. For example:

- I am at work
- I'm in the car

- I'm driving down the A1 and expect to arrive in about 20 minutes
- I'm sitting on the sofa watching television
- I'm with my mother
- The shipment is in our warehouse
- I'm wearing my glasses
- He's in 3rd place, 4 seconds behind John who's leading the race

The interesting thing about all of these examples is that the positional interest is in the relationship between two or more people and/or objects. With reference to Figure 1.1 one could measure the precise latitude, longitude and height of the sofa, the television set and me separately and from this compute the relationships to infer that I was sitting watching television. However, it is a very clumsy (or sophisticated) way of solving the problem, especially as you also need to know that I'm sitting, awake and facing the TV!

The other thing to notice is that in many of the examples above the position information is imprecise and it is the context within which the information is interpreted that matters.

1.4 Location as a context for applications

Whilst sea navigation and some scientific uses of location do require latitude and longitude, many of our day-to-day requirements for position information don't, and as we have seen the value to applications is the context in which the position information is interpreted.

Figure 1.1 Position is about the context

For example if an application requires to know that I am in my car – for example to modify the behaviour and services offered by my mobile phone – it could measure the position of the car and me, compare them and act accordingly. For this the positioning technology would need to work wherever the car goes, including tunnels, multi-storey car parks etc., and it would have to be sufficiently accurate; a non-trivial task. Alternatively a simple proximity detector that works over a few feet and detects when I am sitting in the car could do the same thing more reliably and cheaper.

A system for monitoring the wellbeing and activity level of someone recuperating from an accident is largely unconcerned about exactly where they are, but it is important to keep track of how far and how quickly they've walked, how long they spent sitting down, lying down or standing along with other clinical measurements.

Most of our daily activities involve some aspect of location and position and those applications that make the best use of this information to inform the context and behaviour of the application are likely to be the ones that turn out to be the most useful. Putting it another way – the use of location and position information is a very good way of improving the relevance of the information presented to us by the devices and technology we use in our daily lives.

1.5 Techniques for determining the position of an object

There are many different ways of determining location, so to set the scene this section provides a short description of the different approaches that may be used. The rest of this book focusses on those methods based on the use of modern technology, but references to information about the other techniques is included to allow the reader to widen the scope of their study should they wish.

1.5.1 Observations of the natural world

Using only observations of the natural world it is possible to find one's way using clues from plants, animals, weather, sea, Sun, Moon, stars and nature in general. These techniques are usually imprecise in comparison to modern technological solutions, and very often they rely on direction vectors rather than absolute positions. For those readers wishing to find out more about natural navigation or wayfinding, *The Natural Navigator* [2] [3], is a good starting point.

In a very simple piece of technology, the magnetic compass, we have one of the oldest and most basic instruments for navigation. It is generally believed that the magnetic compass was discovered in China around 200 BC, with very early references to the possible use of magnetic fields for direction finding as far back as 2643 BC.[1] In 1190 Alexander Nekam wrote of sailors using the compass to aid with navigation. One of the earliest descriptions of the principle of operation and construction of the magnetic compass is set out in a letter written by Peter Perigrinus in 1269. Mottelay [4] provides an interesting historical treatise for those interested in reading more of the historical background.

A compass is an instrument that is able to measure and display the direction of the Earth's magnetic field at the point of observation. It is based on the principle that the Earth is like a large magnet and is surrounded by a magnetic field in which the poles of the magnet are located near to (but not coincident with) the poles of the axis about which it rotates. Therefore the direction indicated by the compass, 'magnetic north', differs from the direction of 'true north'. The difference is known as the 'magnetic declination' being positive when magnetic north is east of true north and negative when west of true north. Since the magnetic poles are not co-located with the polar axis of the Earth, the value of the magnetic declination varies depending on one's geographical location. Furthermore the position of the magnetic poles varies with time and this can introduce changes that may drift by several degrees per century. A global spherical harmonic model of the Earth's magnetic field known as the IGRF (International Geomagnetic Reference Field) is maintained under the auspices of the IAGA (International Association of Geomagnetism and Aeronomy). For more detailed information about the Earth's magnetic field a comprehensive starting point is [5].

---

[1] Historians do not believe that there is conclusive evidence of the use of magnetism at this time, although accounts indicate the use of tools that might have made use of magnetism.

Since the Earth's magnetic field is relatively weak it is affected by local variations caused by the presence of magnetic materials, including iron and iron alloys, and electromagnetic fields from electrical machines. This can lead to large local variations. For example using a compass inside a car, which is predominantly built from steel, is notoriously unreliable. Nevertheless, the magnetic compass is still a valuable tool for navigators today, especially considering that modern sensors for measuring magnetic fields, magnetometers, are cheap, accurate and require very little power to operate.

It is interesting to note that most of these natural observation techniques lead to an indication of direction rather than a definitive geographical position. Therefore they need to be linked with observations of specific points, typically geographic features, in order to be useful. The relevance and usefulness of direction and orientation when coupled to a geographic position is an essential part of the art of navigating, locating and positioning and we will return to this theme throughout the book.

1.5.2 Celestial observations

Astronavigation is the art and science of navigation using observations of the Sun, Moon, stars, planets and other celestial bodies. It is one of the oldest practices in human history. Typically navigators use observations of the Sun, Moon or one or more of the navigational stars listed in the nautical almanac.

Polaris, often called the North Star, is one of the best known navigation stars. It lies in a due north direction being less than one degree from the celestial north pole. It is only visible in the northern hemisphere. In the southern hemisphere the Southern Cross constellation is a group of stars that with a simple geometric construction, as illustrated in Figure 1.2, can be used to indicate a true south direction to an accuracy of about 5 degrees.

By making careful observations of the angle between a celestial body and the horizon, or other celestial bodies, and a knowledge of time, it is

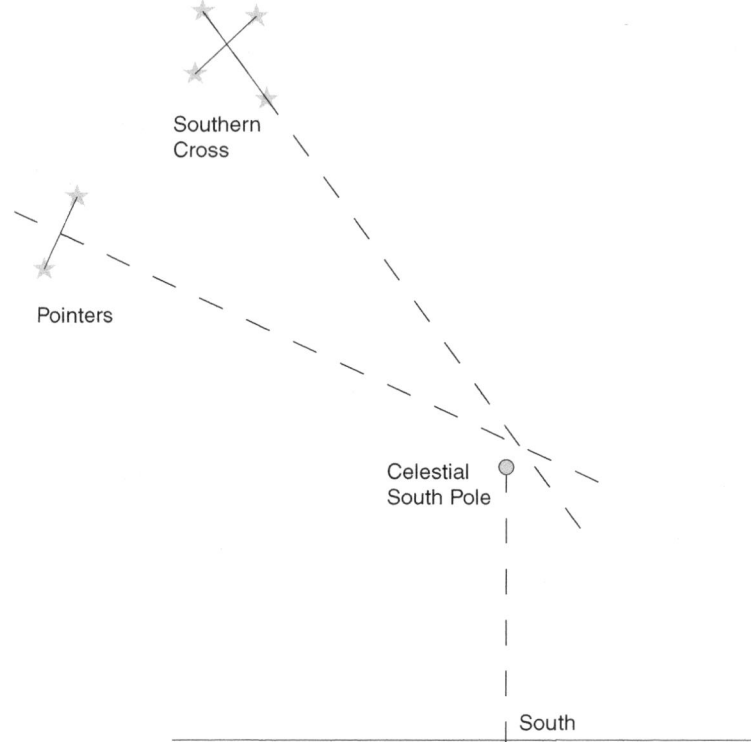

Figure 1.2 Finding South using the Southern Cross and pointers

possible to compute the geographic position of the observer on the surface of the Earth. The sextant has long been the instrument used by sailors to determine their positions at sea. It provides an accurate measurement of the angle between two objects by sighting both simultaneously. An excellent general reference for navigation including a chapter on the sextant is Bowditch [6] [7].

For more in-depth information about celestial navigation, the reader is referred to Karl [8], though there are many books on the subject. No further references will be made to celestial navigation in this book.

The role of time, and having an accurate clock was one of the technological limitations that plagued early navigators. Sobel [9] provides an interesting and absorbing read about John Harrison and his lifetime spent developing precision mechanical clocks in the eighteenth century, and the role they played in ocean navigation. Time is an absolutely crucial element of many navigation techniques, and the use and role of time will also be a recurring theme throughout this book.

1.5.3 Using radio signals

There are numerous systems and techniques for using radio signals to aid in the determination of the position of an object. They can be broadly divided in three classifications:

1. Direction finding;
2. Range measurement, generally using Time of Flight, Time of Arrival, Time Difference of Arrival, signal strength or other characteristics of the radio signal;
3. Velocity measurement, generally based on the principles of Doppler shift.

The earliest systems used direction finding by rotating the antenna to find the strongest signal and hence the direction. These techniques are still used in simple tracking systems such as many used in conservation for wild animal tracking.

Later the principle of a transmitted signal received at two different locations, or two different synchronised transmitters received at one location was used to determine the difference between the times of arrival of the two signals and this led to the category of hyperbolic systems, which given two pairs of signals are able to determine a unique position (two-dimensional) for the transmitter or receiver. One of the earliest such systems was DECCA. A more recent example of such a system is LORAN, which operated until 2010.

The same techniques, with some adaptation, can make use of public signals, such as those from mobile cellular network base stations, or signals transmitted by TV and radio broadcast transmitters.

GPS is another system that uses radio signals, this time transmitted by a constellation of satellites orbiting the Earth. A GPS receiver measures the times of arrival of signals from multiple satellites in order to determine the position of the receiver. It may also measure and make use of signal Dopplers to determine velocity.

Whilst most modern systems rely on an infrastructure of either transmitters or receivers at known locations in order to track the position of a mobile object, there is significant effort going into the development of new architectures for collaborative positioning in which measurements between mobile devices can be used to determine their relative positions without recourse to fixed infrastructure.

A large part of this book will be devoted to more detailed description and analysis of modern locating and positioning systems using radio signals. For information about the history of radio locating systems, or for information about historical systems, the reader is referred to some of the excellent reference works on the subject.

1.5.4 Inertial techniques

According to Newton's first law of motion: '*Every body remains in a state of rest or uniform motion (constant velocity) unless it is acted upon by an external unbalanced force. This means that in the absence of a non-zero net force, the centre of mass of a body either remains at rest, or moves at a constant speed in a straight line.*' Inertia is the property of a body that links force with mass and change of velocity and is the physical basis for inertial navigation.

An inertial coordinate frame is one in which Newton's laws of motion hold true: they are neither rotating nor accelerating. This is usually not the same as a navigation coordinate system which may be fixed to the surface of the Earth – which is rotating and translating. Chapter 2 introduces coordinate systems and explains how they can be related to one another.

Inertial navigation is the principle of using sensors to measure changes in the equilibrium state of Newton's laws of motion and using the measurements to compute the new position (including velocity, acceleration and attitude) of the body within a coordinate system at a later date. The main sensors used for inertial navigation are: accelerometers and rate gyroscopes.

An accelerometer measures rates of acceleration (change of velocity caused by the force acting on the body), which, integrated once gives velocity and integrated a second time yields position information – both assuming that the initial conditions of position and velocity at the start of the integration time are known or can be determined.

A rate gyroscope measures rate of rotation and is used to keep track of the orientation of the object within the coordinate frame. Integrating the rate of rotation yields an orientation (direction), assuming that the initial conditions are known or can be determined.

An inertial navigation system uses the measurements from accelerometers and rate gyroscopes in order to compute position, velocity and orientation over time. Chapter 5 presents a more detailed introduction to the principles of inertial navigation techniques.

1.5.5 Contextual

As human beings we naturally navigate by context. For example when we give directions to someone they are typically something like: 'take the A10 out of town; after about 2 miles turn left at the signpost to *someplace*; follow the road across two roundabouts and at the Barley Mow pub turn right; go past the church and we are third on the left with a green fence in front.'

Even for local positioning we use contextual information as our primary means for locating things: 'your glasses are on the coffee table next to the sofa in the TV room'.

Of course these methods require the ability to extract contextual information from our environment to recognise and link objects within it. Being able to associate position and location information with our environment in this way is one of the big challenges and also one of the major opportunities for locating and positioning systems in the future.

Modern car navigation systems are able to present navigation information to us in this way by associating location with map data and road features that we recognise. Even though based on GPS, they rely heavily on map matching techniques – the 'satnav' assumes that one is driving on a road and 'snaps' the car position to the appropriate road using map matching techniques. For those situations when satellite coverage is lost, such as in tunnels and some urban areas, the 'satnav' may use auxiliary data such as the odometer (wheel counting) input to indicate distance travelled and from that to infer position.

Radio navigation systems may also use contextual information such as proximity to identified radio transmitters, such as Wi-Fi access points, radio or TV transmitters or cellular base stations.

The use of optical techniques for scene recognition is also emerging with the widespread availability of low-cost cameras and imaging sensors. Optical techniques can also be used in specialist areas, such as for target tracking of specific objects – such as the ball used in sports like tennis, cricket and football.

Chapter 5 takes a further look at the various auxiliary techniques, and in Chapter 6 we show how they can be combined to give the best of multiple sensory inputs and in this way move towards the holy grail of ubiquitous positioning capability.

In Chapter 9 we look at applications for positioning and location, especially considering what lies beyond present navigation systems.

2   Coordinate systems

The two most common coordinate systems used for positioning and navigation are:

1. Latitude, longitude and height above mean sea level;
2. Cartesian systems in which (x,y,z) axes are arranged orthogonally to one another.

However, in our day-to-day lives, most position and location information is described contextually: I'm at work; I'm sitting at my desk; I'm in the car; I'm travelling south on the M1; I'm sitting next to Jim; etc. This sort of contextual information is what most applications would ideally like to have available to them, but for the purposes of computing a position or location, most locating systems work in Cartesian space (including GPS), although the resulting position may be presented as latitude, longitude and height.

2.1 Latitude and longitude

2.1.1 A simple definition

Latitude and Longitude are two angles used to describe a terrestrial location on the surface of the Earth. See Figure 2.1.

    Latitude is measured north and south from the equator, which is an imaginary line running around the circumference of the Earth so that 90 degrees north is the north pole and 90 degrees south is the south pole. Lines of latitude run east–west linking points having the same angular measurement from the zero degree parallel, the equator.

    Longitude is measured east and west from the prime meridian, some-times called the Greenwich meridian because it was defined to run through the Royal Observatory, Greenwich, in the United Kingdom.

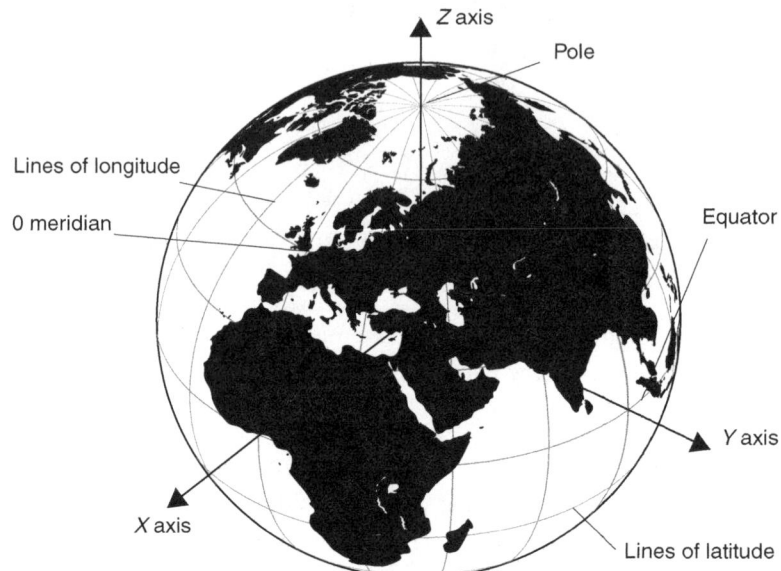

Figure 2.1 Latitude and longitude and ECEF Cartesian axes

Longitude runs from 0 to 180 degrees east and 0 to 180 degrees west, with the 180 degrees east and west lines being the same meridian on the opposite side of the globe. Lines of longitude run north–south linking points having the same angular measurement from the zero meridian. All lines of longitude run through the north and south poles where they converge.

2.1.2 But it's not as simple as this

Unfortunately whilst conceptually simple, latitude and longitude as a measure of location need to be used carefully and wisely if one is interested in high precision.

- There are many different definitions for the equator and prime meridian, leading to the confusing situation that the same coordinates in different systems of latitude and longitude may differ significantly (hundreds of metres). Or put the other way, the same point on the ground may have significantly different coordinates in the different systems. There are still systems in existence in which a different prime meridian is chosen, for example, French Institut Géographique National (IGN) maps use a prime meridian passing through Paris, although they also include measurements from the Greenwich meridian. It is interesting to note that it is as recently as 1884 that near universal agreement to use the same prime meridian was reached.

- The Earth is a spheroid – a sphere-like shape – often called a geoid. It can be approximated mathematically as a biaxial ellipsoid, but being irregular in shape this approximation is at best just that, an approximation. Measures of longitude and latitude are based on a mathematical approximation to the shape of the Earth. Many different models are used today – chosen for very good, but different, reasons. This topic is covered in more detail later in this chapter.

- The Earth is a continuously changing and evolving object. Points on the surface (identifiable features) are moving relative to one another. Long-term drift is sometimes referred to as continental drift which, in some parts of the world, can be as much as 10 centimetres per year. On a shorter timescale the surface of the Earth rises and falls daily and annually on a millimetre scale with tidal and weather variations and with the seasons, such as freeze and thaw of the ice shelf.

Therefore if one plans to use a global coordinate system for local applications, it is important to realise that everything may not be quite what it seems.

2.1.3 Systems of latitude and longitude

With the widespread use of GPS there has been real convergence around a single global Terrestrial Reference Frame (TRF) called WGS84 or GRS80 or ITRS (essentially the same although they are very slightly different) for latitude, longitude and height. However, most countries or regions also have a local reference frame that provides a more accurate representation of the part of the Earth of interest. The other significant difference between the global and local TRFs is that the global TRF is regularly updated based on a wide selection of observed reference sites around the world, and it therefore represents the 'mean' position, whereas local TRFs are aligned to local features anchored on the tectonic plate on which the region is located, and therefore local TRFs are moving relative to the global TRF. The differences between them can be quite large, but since tectonic plate movement is steady and predictable it is relatively easy to relate them to one another based on the time of the location observation.

2.1.4 The global TRF

There is actually no such thing as a single global TRF. However, there is a clear trend towards harmonising the three main contenders:

- WGS84 (World Geodetic System of 1984) is the reference frame used by the United States of America Department of Defense and is the basis of the GPS broadcast and 'precise' orbits. It is updated from time to time and is now closely aligned with ITRS 94 (the IERS terrestrial reference frame – IERS is the International Earth Rotation Service). GPS receivers usually output their positions using WGS84.
- GRS80 (Geodetic Reference System 1980) is a reference ellipsoid developed to closely match the Earth as a whole. It is an ellipsoid designed to approximate the mean sea level across the entire globe under ideal, calm, non-perturbing conditions. In practice the Earth is a somewhat irregular geoid and the ellipsoid varies from the actual surface by up to 110 metres. GRS80 and WGS84 differ by very small amounts (only a few millimetres), and are, for most practical purposes, the same.
- ITRS is the reference frame of the International Earth Rotation Service. It and WG84 have now been harmonised to the extent that they are practically the same. However, the ITRS is based on measurements from more than 500 stations around the world with four different space positioning methods being used: Very Long Baseline Interferometry (VLBI), Satellite Laser Ranging (SLR), GPS and Doppler Ranging Integrated on Satellite (DORIS). This makes it the most accurate TRF constructed, even better than the military WGS84 TRF.

The size and shape of the ellipsoid is fully defined by two parameters, the semi-major (a) and semi-minor (b) axes. For GRS80 these parameters (*metres*) are:

$$a = 6378137.0000$$
$$b = 6356752.3141 \cdot$$

(2.1)

From these we can derive a single parameter used to describe the shape of the ellipsoid: the ellipsoid squared eccentricity e^2. Sometimes it is useful to define the eccentricity in terms of a flattening f:

$$e^2 = \frac{a^2 - b^2}{a^2} \cdot$$
$$e^2 = 2f - f^2$$

(2.2)

Many other ellipsoid models for the Earth have existed and been used in the past, and many are still in use for legacy and other long-standing products. However, for the purposes of this book and modern practitioners we will not consider them any further.

Using modern survey techniques and instruments the TRF is kept aligned to the mean of a large number of fixed reference sites across the entire planet (Figure 2.2). Therefore local features gradually move over time relative to the TRF due to continental drift.

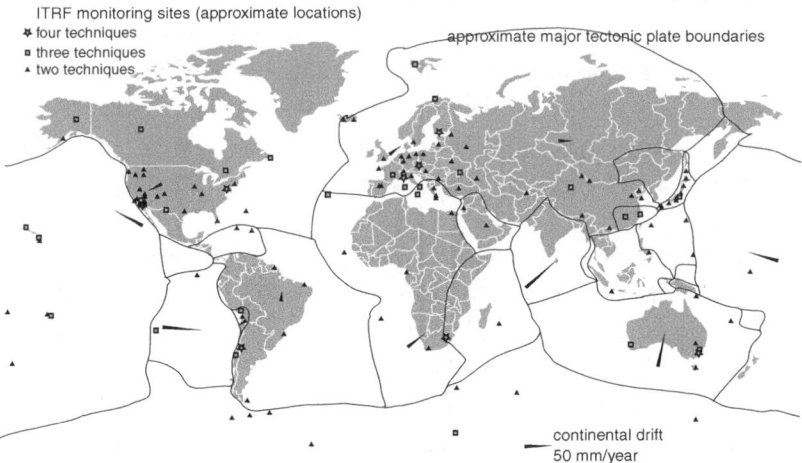

Figure 2.2 Map showing ITRF monitoring stations overlaid with tectonic plate boundaries and vectors indicating continental drift (ITRF data 2005)

2.1.5 TRFs for regional use

There are several reasons why countries or regions use a different TRF for local and high-resolution maps of the area:

- A different ellipsoid model can provide a much more accurate fit over the smaller region of interest. This is particularly true for height.
- The TRF is aligned to local features of the region and therefore moves with the drift of the tectonic plate on which the region is based. This means that local features remain at the same coordinates for much longer periods of time, and there is no need for most applications to account for continental drift.
- There may be legacy products and maps or data based on an older TRF and the cost and complexity of changing doesn't warrant the benefit.
- Coordinates may be presented using UTM coordinates (often a localised variant of UTM) rather than as latitude and longitude.

There are many different local TRFs in use, for example:

- North America uses NAD83 (North American Datum 1983) and NAD27 (North American Datum of 1927), the latter being phased out gradually. NAD83 is based on GRS80 but is linked to geographical features of the region and it is therefore drifting relative to WGS84, but very slowly. Since being established they have drifted apart by about 1 metre. The biggest difference is in height with differences ranging from a few metres in some areas in the east to almost 100 metres in the west. NAD27 is based on the Clarke 1866 ellipsoid with a reference point at Meades Ranch in Kansas. The National Geodetic Survey website has plenty of information and further details for readers needing to find out more.
- A TRF has been defined for Europe, ETRS89 (European Terrestrial Reference System 1989). In 1989 it coincided with WGS84, but is drifting with the European land mass, at around a few centimetres per year. The UK has adopted ETRS89 for the 'OS Net' products, but the UK 'National Grid' is based on a TRF called OSGB36 (Ordnance Survey Great Britain) which uses the Airy ellipsoid. For more information about UK mapping and systems the OSGB paper [10] provides an excellent introduction.

There are too many local TRFs for different regions to describe them all here. The reader is referred to their local mapping or government survey office for more information.

2.1.6 So what about height?

So far we've discussed position on the surface of the Earth as a latitude and longitude but, apart from noting that the TRF ellipsoid differs in height because of the wrinkly surface of the Earth, have not said anything else about height.

The reference surface for the purposes of physical height measurement is referred to as Mean Sea Level (MSL). Heights above are positive and those below negative. This surface is generally described as 'level' by which we mean that it is perpendicular to the gravity vector. Since gravity has many minor variations from place to place (since the mass of the planet is not evenly distributed throughout), this surface called mean sea level is actually lumpy and uneven. When we describe one point as being higher than another we mean that the level surface passing through the lower lies below the higher point.

Systems like GPS measure ellipsoid height, which, without knowledge of the deviation of the actual geoid from the TRF ellipsoid (Figure 2.3), is sometimes only useful for making relative height measurements.

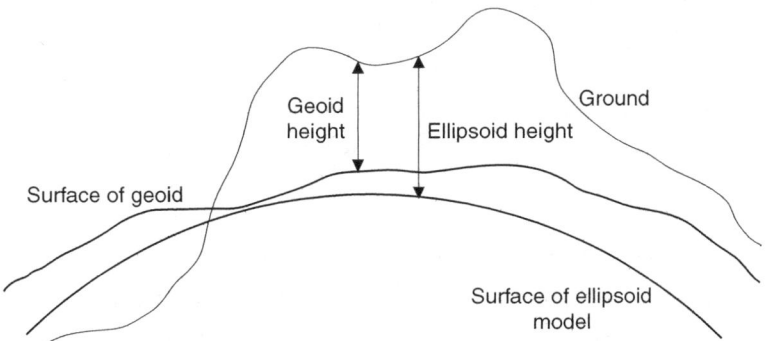

Figure 2.3 Illustration of geoid and ellipsoid heights

Whereas latitude and longitude using a global or local TRF are easily related to one another the task of relating height to the true MSL height for an arbitrary location near the surface of the Earth is far from trivial. Most countries have, therefore, adopted their own reference basis for height.

In the USA, a vertical control datum called NGVD88 (National Geodetic Vertical Datum 1988) has been established to supersede NGVD29 (1929) as a basis for MSL height across North America.

In the UK a reference datum called ODN (Ordnance Datum Newlyn) has been established in which the mean sea level measured at Newlyn Cornwall between 1915 and 1921 has been adopted as the zero height reference level. Comprehensive measurements across the entire area of the British Isles have been used to define MSL heights across the whole region. This has led to the development of the National Geoid Model OSGM02, which may be used to relate ellipsoid height (from GPS) to actual height above MSL for anywhere across the British Isles.

2.2 Cartesian coordinate systems

Rectangular Cartesian coordinates are the most common and widely used method for representing positional information. For a three-dimensional solution three perpendicular axes, which are usually called X, Y and Z, are used. They are usually arranged in a right-handed configuration. Inverting any one axis results in a left-handed arrangement, but this book assumes right-handed axes throughout.

In order to have a practically useful Cartesian coordinate reference frame it is necessary to define five initial conditions: The position of the origin (X_0, Y_0, Z_0) and the orientations (directions) of two of the axes.

2.2.1 Cartesian coordinates for global positioning

For global positioning in the context of the TRFs described in the previous section of this book it is useful to define a Cartesian coordinate system having the origin at the centre of the ellipsoid, the X axis lying on the equator passing through the prime meridian and the Z axis coinciding with the polar axis of the ellipsoid the north pole being positive. As a result the Y axis also lies on the equator passing through the 90° east meridian. See Figure 2.1.

This is commonly referred to as Earth Centred Earth Fixed (ECEF), and is the coordinate frame usually used to compute GPS positions within the GPS computation engine, the result being converted to latitude, longitude and height before being output by the receiver.

As a result any latitude and longitude can be expressed as a unique (x, y, z) set of coordinates. Given latitude ϕ, longitude λ and ellipsoid height H, plus major axis of the ellipsoid a and ellipsoid eccentricity e:

$$
\begin{aligned}
v &= \frac{a}{\sqrt{1 - e^2 \sin^2 \phi}} \\
x &= (v + H) \cos \phi \cos \lambda \\
y &= (v + H) \cos \phi \sin \lambda \\
z &= \left((1 - e^2) v + H \right) \sin \phi
\end{aligned}
\tag{2.3}
$$

Converting from (x,y,z) to latitude ϕ, longitude λ and ellipsoid height H is also straightforward, but it requires an iterative implementation.

First longitude is easily computed:

$$
\lambda = \arctan \left(\frac{y}{x} \right).
\tag{2.4}
$$

Next compute an initial value of latitude:

$$
\phi = \arctan \left(\frac{z}{\sqrt{x^2 + y^2 (1 - e^2)}} \right)
\tag{2.5}
$$

then iteratively improve the value of latitude until sufficient precision has been reached:

$$
\begin{aligned}
v &= \frac{a}{\sqrt{1 - e^2 \sin^2 \phi}} \\
\phi &= \arctan \left(\frac{z + e^2 v \sin \phi}{\sqrt{x^2 + y^2}} \right)
\end{aligned}
\tag{2.6}
$$

and finally compute the ellipsoid height:

$$H = \frac{\sqrt{x^2 + y^2}}{\cos \phi} - \nu. \qquad (2.7)$$

2.2.2 Transverse Mercator map projections

Since the surface of the Earth is curved, when it is drawn as a map on a flat piece of paper it has to be warped, or distorted into a rectangular shape. The proof that this is true is contained in Carl Friedrich Gauss's *Theorema Egregium* [11]. The function that converts ellipsoidal longitude and latitude coordinates to plane coordinates is called the map projection.

For most applications this form of representation in which we have planar coordinates on a horizontal plane on the surface of the Earth is most convenient. In reality the surface of the Earth is curved, but depending on the accuracy required and the importance of measurements such as distance, area and direction to the application, a flat representation is fine for limited areas of the Earth's surface.

There are many different map projections, each useful for a different purpose or for a different scale of mapping. Generally it is possible to preserve one or some of the following characteristics, but never all: area, shape, direction, bearing, scale, distance. It is beyond the scope of this book to go into details of the different projections and the compromises they make, so the reader is referred to a reference such as Snyder [12], for further information. However, it will be useful to say a little about UTM (Universal Transverse Mercator).

The UTM system divides the surface of the Earth into 60 zones each 6° longitude in width, centred on a longitude meridian and extending from 80° south to 84° north. The zones are numbered from the west starting with 1 up to 60 with 30 immediately west (3°) of the Greenwich meridian and 31 immediately east (3°) of the Greenwich meridian.

Each of the zones is mapped using a transverse Mercator projection, based on the WGS84 ellipsoid, which preserves north–south distances along the centre of the zone, but distorts east–west distances. Since each zone is narrow (6°) the amount of east–west distortion is quite small. A scale factor of 0.9996 is applied along the centre which leads to a scale

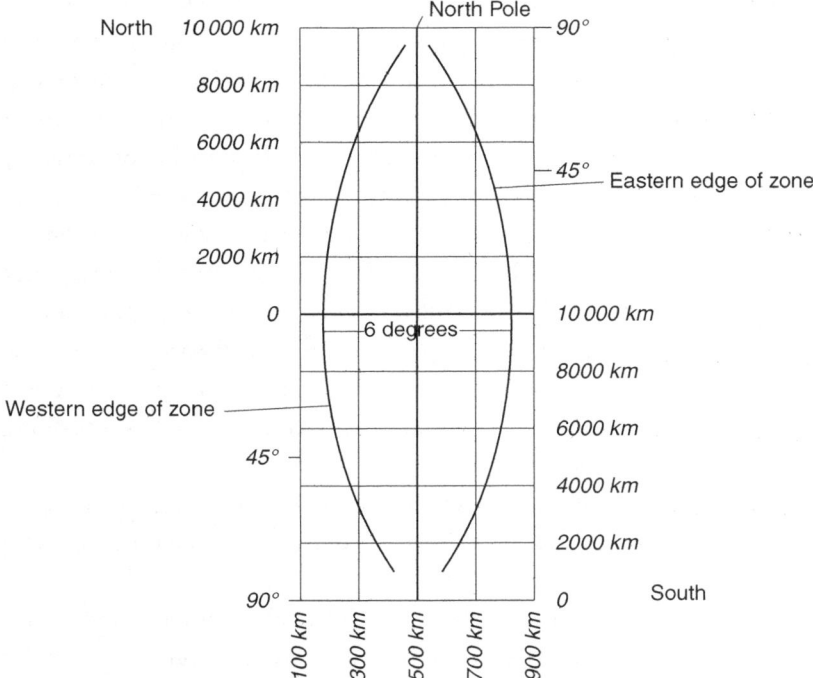

Figure 2.4 Illustration of a UTM zone

factor of about 1.001 at the outer edges of the zone at the equator. Therefore the scale error is held below 1 part in 1000 across the entire zone. See Figure 2.4.

A position on Earth consists of its UTM zone reference and an easting and northing coordinate pair. The easting is the distance east from the centre line of the zone, while the northing is the projected distance from the equator. In order to avoid the use of negative numbers the easting is offset by a false easting value of 500 000. Therefore eastings range from 167 000 to 833 000 metres. Northings are offset by a false northing value of 10 000 000. Coordinates in UTM are given in metres.

The above is a simplified description of UTM; for more information about UTM, the reader is referred to the Military map reading guide [13].

In addition to the globally defined UTM coordinate system, there are also a number of geographically local definitions based on the transverse Mercator projection. One example is the national grid used in the United Kingdom administered by Ordnance Survey. The projection used has a central meridian of 2° west and a central scale factor of approximately 0.9996. The true origin is at 49° north and 2° west. A false origin at 400 km west and 100 km north is defined so that all coordinates on the British Isles are positive. The area is divided into uniform 100 km squares and two letters are used to unambiguously describe each square. Numeric easting and northing coordinate pairs are used to describe the precise location within each square. So, for example, the grid reference TL455581 defines a point in grid square TL with easting 455 and northing 581, which is a 100 m by 100 m square.

Other countries that have defined national grid systems include: Australia, Belgium, Finland, Ireland, Italy, The Netherlands, New Zealand and others.

Coordinate conversions between UTM and ellipsoid coordinates, whether latitude and longitude or Cartesian, are not trivial, and require the ellipsoid parameters to be taken into account. The most straightforward conversion is between coordinates based on the same TRF; for example NAD83 coordinates (GRS80) and GPS latitude and longitude (WGS84 which is the same as GRS80).

The British National Grid uses the OSGB36 TRF which is based on the Airy ellipsoid. In this case the transformation between GPS coordinates (ECEF or WGS84) and OSGB coordinates is more complex because not only is the coordinate transformation required, but also a conversion between different datums. The reader is referred to Ordnance Survey for more detailed information.

Furthermore if one needs to convert height values, such as the OSGB height above MSL as used on OS maps, or the NAVD geoid height used by National Geodetic Survey of America, to or from ellipsoidal height such as measured by GPS it is necessary to use the geoid height models adopted by the relevant system. The reader is referred to the respective national bodies for further information and access to geoid model parameters.

2.2.3 Arbitrary Cartesian coordinate systems

The UTM coordinate system described above is a three-dimensional Cartesian system if one includes height information. In this case one can view the easting as the X axis, the northing as the Y axis and height is the Z axis with the origin of the coordinate system at the false origin of the UTM zone being used. Unfortunately this leads to different interpretations of angles in a mathematical sense and geographical compass bearing sense. We will come back to this later on in the discussion of directions and review other options for dealing with angles. Height is generally understood to increase upwards with the gravity vector pointing down.

However, for many applications this is over-complicated and it may be better to use an arbitrary locally defined coordinate frame. For example, it may be convenient to define a local coordinate system for a sports field when tracking players around the field, or to define a local coordinate system for a building in order to track goods and people around the premises.

In order to associate this frame with the environment it is necessary to define the position of the origin and the directions of two of the axes. Transformation between different Cartesian systems is easily achieved with knowledge of their relative offset and rotation.

2.3 Inertial coordinate frame

The description in the previous section has been entirely based on the assumption that the coordinate reference frame we are using is associated with the physical world in which we live, and indeed this is the most useful representation in the context of our sensory perception of the environment and the way in which we navigate and perceive the world around us.

However there is a special case for a different reference frame, and that is a coordinate system fixed in inertial space. According to Newton's laws of mechanics the motion of a body will continue uniformly in a straight line unless disturbed by an external force acting on the body. A given force will produce a constant acceleration. These properties also apply to the rotation of the body.

Since the Earth rotates on its axis and travels around the Sun, any reference frame attached to features of the Earth is subjected to changes in orientation and velocity. For inertial navigation it is necessary to define a reference frame that is invariant in time and space, and is, therefore, not attached to the Earth.

Inertial navigation techniques are introduced in Chapter 5 at which point we'll return to the question of defining and using an inertial reference frame.

2.4 Describing direction and orientation

2.4.1 Simple direction on a plane surface (e.g. map)

In a geographic sense for location on the surface of the Earth direction is usually described using a compass bearing. Compass bearings are usually in degrees and rotate **clockwise** from north which is 0°. Therefore east is 90°, south 180° and west 270°. Compare this to the traditional way of measuring angles mathematically in two dimensions: traditionally the x axis points right and the y axis up (forwards) with angles measured from the x axis in an **anticlockwise** direction. See Figure 2.5. In both of these cases z is up or out of the page.

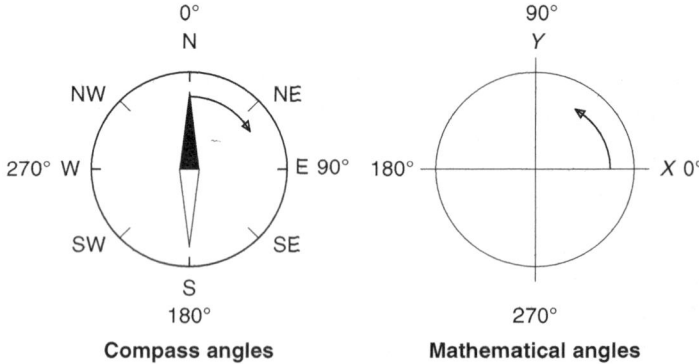

Figure 2.5 Comparison of compass and mathematical angles

Therefore we have two choices:

- We can adopt a coordinate system in which north aligns with the positive x axis, east with the positive y axis and mathematical angles correspond to compass bearings, but the z axis (height) is positive down (sometimes referred to as an NED reference frame) which is the opposite polarity to that used for height in a traditional geographic sense; or
- We can keep height as up and live with the disparity between conventional compass bearings and mathematical angles which start at a different point and run in opposite directions.

It is common in location calculations to use the former definition, especially for reference frames attached to a moving object. We will come back to this in the chapter covering inertial navigation.

Whichever coordinate frame is used in a planar two-dimensional sense direction can be described using a single measure, such as the compass bearing.

2.4.2 Direction in three dimensions

Extending this to three dimensions one can define a direction based on two angular measurements: azimuth (bearing) and elevation (angle above or below the horizon). Whilst this fully describes the direction it is an incomplete description of orientation because it does not say anything about rotation around the direction vector.

Therefore direction may not fully describe orientation, but knowledge of orientation can tell us direction (Figure 2.6).

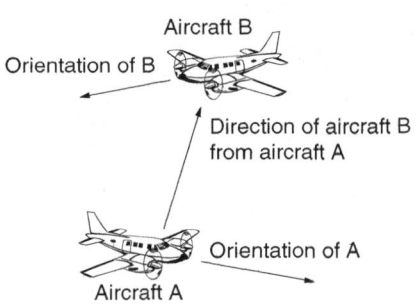

Figure 2.6 Orientation and direction compared

2.4.3 Describing orientation: Euler angles

Very often the term attitude is used instead of orientation, particularly for airborne vehicles, but since we are taking a much wider view of what the object being tracked is, we will use the more general term orientation.

A common way of describing orientation is with three angles, often described as yaw, pitch and roll. These three angles are called the Euler angles and are conventionally defined as follows:

ψ (yaw): rotation about the z axis
θ (pitch): rotation about the y axis
ϕ (roll): rotation about the x axis

The axes in this case are the body axes of the object whose orientation we are describing and the angles describe the orientation of the object's body axes with respect to the navigation coordinate frame within which we are operating. They are fixed to the object and are conventionally aligned with the x axis pointing forwards, the y axis to the right and the z axis pointing down (Figure 2.7).

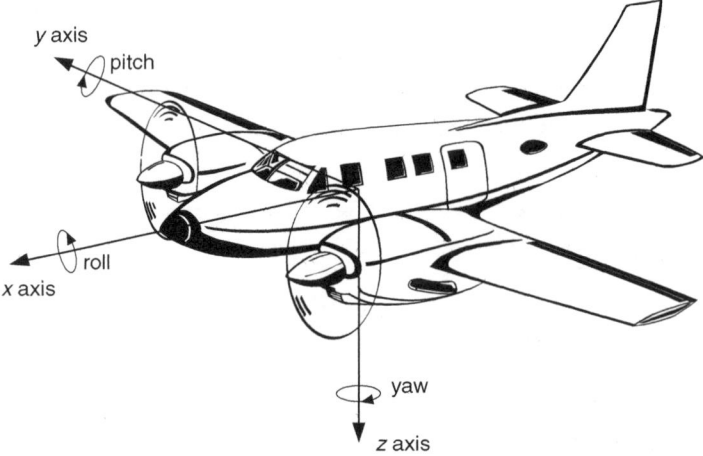

Figure 2.7 Body axes defining orientation of an object

The three rotation matrices are:

$$\mathbf{R}_x(\phi) = \begin{bmatrix} 1 & 0 & 0 \\ 0 & \cos\phi & \sin\phi \\ 0 & -\sin\phi & \cos\phi \end{bmatrix} \tag{2.8}$$

$$\mathbf{R}_y(\theta) = \begin{bmatrix} \cos\theta & 0 & -\sin\theta \\ 0 & 1 & 0 \\ \sin\theta & 0 & \cos\theta \end{bmatrix} \tag{2.9}$$

$$\mathbf{R}_z(\psi) = \begin{bmatrix} \cos\psi & \sin\psi & 0 \\ -\sin\psi & \cos\psi & 0 \\ 0 & 0 & 1 \end{bmatrix}. \tag{2.10}$$

It is important to realise that as a rotation is applied to the object the body axes remain fixed to it and the rotation we are referring to is the rotation of the body axes with respect to the navigation reference frame which we are working in. This is, therefore, a transformation between two axis reference frames and not the rotation of a vector within a coordinate system – the Euler angles actually represent three successive rotations about the three axes. Euler angles are only unambiguous if we are diligent about the order in which they are applied. By convention the angles are defined in the order: yaw, ψ (rotation about z axis), pitch, θ (rotation about y axis), roll, ϕ *(rotation about x axis) to move from the navigation frame to the body frame. To 'unwind' the body frame into the navigation frame the inverse sequence must be used: roll, then pitch, then yaw.*

Whilst Euler angles represent a simple way of describing the orientation of one reference frame with respect to another using the minimum set of information, they are not usually the most convenient mathematical representation for use in navigation processing systems. Two alternative representations of orientation are generally preferred:

Direction Cosines
Quaternions

Each of these will be described briefly – for more detail the reader is referred to some of the excellent references on the subject.

2.4.4 Direction cosines

In three-dimensional space direction cosines comprise a 3×3 matrix in which the columns are unit vectors representing each of the body axes projected onto the navigation frame axes which are represented by the rows. The name arises because each element c_{ij} (ith row and jth column) is given by the cosine of the angle between the i axis of the navigation frame and the j axis of the body frame

$$\mathbf{C}_b^n = \begin{pmatrix} c_{11} & c_{12} & c_{13} \\ c_{21} & c_{22} & c_{23} \\ c_{31} & c_{32} & c_{33} \end{pmatrix}. \tag{2.11}$$

2.4.5 Quarternions

Quarternions form a class of algebra rather like an extended set of complex numbers taking the general form:

$$q = a + ib + jc + kd \tag{2.12}$$

in which i, j and k represent imaginary numbers where

$$i = j = k = \sqrt{-1}. \tag{2.13}$$

This four-dimensional representation was discovered by William Hamilton in 1843 and one of its main applications is in algebra to represent 3D spatial transformations of reflection, rotation and scaling but not translation.

Writing the quarternion above as:

$$q = a + \hat{u} \tag{2.14}$$

it can be shown that the unit quarternion:

$$q = \cos \theta + \hat{u} \sin \theta \tag{2.15}$$

represents the rotation of a 3D vector by an angle 2θ about the 3D axis u.

For more information and a detailed treatment of the subject the reader is referred to one of the specialist publications such as the book by Kuipers [14].

3   Satellite positioning (GNSS)

3.1 Introduction to satellite positioning

At the time of writing there are two operating Global Navigation Satellite Systems (GNSS). The best known of the satellite positioning systems is GPS (Global Positioning System) which is the satellite-based navigation system developed by the US Department of Defense under the NAVSTAR programme. The other is GLONASS, the Russian system. Although it had fallen into disrepair it is now restored to full operational status.

There are additional planned systems, including GALILEO, COMPASS and others. Galileo and Compass are both in their pre-operational phases and are expected to become operational in the time-frame 2014 to 2018.

GNSSs allow a receiver which can receive radio signals from four or more navigation satellites (in general) to compute its position within an Earth fixed reference coordinate frame such as WGS84. The key features are:

- It is a one-way system in which the satellites transmit navigation signals and receivers receive the signals and use them to compute their own positions (usually).
- The receiver must be able to receive signals from four or more satellites (for normal 3D positioning), and therefore it does not work in places where the signals cannot be received. The better the quality of the signals received, the better the positioning.
- The quality of the position fix is affected by the environment in which the receiver is operating: multipath, diffraction and interference all affect performance.
- The receiver needs to know the positions of the satellites transmitting the signals; this information is broadcast by the satellites and usually decoded from the received signals by the receiver.

- The position and velocity may be computed from received signal measurements, but not orientation (at least not without special receiver architectures or additional sensors).
- The resulting position is usually presented as latitude, longitude and ellipsoid height in a coordinate space defined by an ellipsoid model of the Earth.

This book only aims to introduce GNSS, with emphasis on GPS, it being the most widely used system today, highlighting some of the important considerations for anyone intending to use GNSS, to the level that they have an understanding of its capabilities and limitations. For more details the reader is referred to some of the excellent and comprehensive texts on the subject.

3.2 Brief description of how GPS works

3.2.1 Simple overview

1. A GPS receiver receives signals transmitted by four or more (usually) satellites which are part of a constellation orbiting the Earth (Figure 3.1).
2. It measures the precise time of arrival of the signals.
3. The difference in time between when the signal was transmitted and when it was received indicates the distance that the signal has travelled.
4. It extracts data messages from the satellite signals. This information tells it where the satellites are in space-time.
5. Using a process of trilateration the receiver is able to compute its own position and also the difference between its clock and the clock used by the satellites.
6. It also applies corrections to compensate for known errors and signal propagation characteristics through the atmosphere.

3.2.2 Satellite infrastructure and architecture

GPS comprises three main segments: Space Segment, Control Segment and User Segment. The US Air Force develops, maintains and operates the space and control segments. See reference [15] for details of the space segment and navigation user interfaces.

Figure 3.1 GPS Constellation (US National Oceanic and Atmospheric Administration)

The Space Segment consists of the orbiting satellites – called Space Vehicles (SV). They are arranged to orbit the Earth on 6 orbital planes, each having at least 4 satellites, with the total number of orbiting satellites being between 24 and 32. The satellites operate in medium Earth orbit (MEO) at approximately 20 200 km above the surface of the Earth. Each satellite circles the Earth in half a sidereal day.[1]

---

[1] A sidereal day is the time it takes for the Earth to make one complete revolution on its axis relative to an inertial reference frame aligned with the stars. It is slightly less than a solar day because the Earth is orbiting around the Sun and therefore has to complete slightly more than one revolution in order for the same point on the surface to be facing the Sun each solar day. A sidereal day is approximately 23 hours, 56 minutes and 4.1 seconds.

Since there are 6 orbital planes each plane is separated by 60 degrees around the Earth. Each plane is tilted at about 55 degrees inclination. The positions of the satellites in each orbital plane are arranged so that at least 6 satellites are within view everywhere on the surface of the Earth. This means that they are, therefore, not spaced evenly around each orbit.

At present (July 2012) there are 32 GPS satellites in orbit (Figure 3.2), one of which is presently out of service. From time to time satellites are taken out of service for maintenance or orbital corrections; generally there are at least 29 usable satellites operating at any one time.

The Control Segment is the part of the system used to monitor and control the satellites. It consists of master and backup control stations, and a number of dedicated monitoring stations. Measurement and tracking information for each satellite is processed by the US Air Force 2nd Space Operations Squadron which regularly contacts each satellite to send it navigational updates and corrections. These updates synchronise the satellite clocks to within a few nanoseconds, and adjust the ephemeris for the satellite's orbital model.

From time to time a satellite needs to be repositioned. It is first marked as unhealthy, then the manoeuvre is carried out and only once the ground

Figure 3.2 Block IIR(M) satellite (GPS.gov library image)

tracking stations have confirmed and verified the new orbit is the new ephemeris uploaded and it is marked healthy again.

3.2.3 GPS signals

The current and legacy GPS system transmits on two frequencies: L1 (1.57542 GHz, which is 154 × 10.23 MHz) and L2 (1.22760 GHz, which is 120 × 10.23 MHz). The L1 band is used for civil signals (the C/A code, originally intended for coarse acquisition, hence its name) and L2 is used for military signals (P(Y) code, precision code). However, GPS modernisation is under way with new signals and bands being introduced. A few satellites are already transmitting a new civil signal in the L2 band and some satellites are transmitting a new military signal in the L1 band. A new band called L5 (1.17645 GHz) is being introduced and the first GPS satellite transmitting an L5 signal was launched in May 2010, and the second on 16 July 2011.

A further signal called L1C that uses a Multiplexed Binary Offset Carrier (MBOC) modulation scheme will be added in the future, starting from around 2014 with full capability anticipated for 2021. This signal has been developed as a common signal used by Galileo and GPS, and it is also being adopted by Japan for the QZSS (Quasi-Zenith Satellite System), China's Compass system and the Indian Regional Navigation Satellite System (IRNSS).

The L1 signal transmit power is approximately 45 W. It is transmitted as a right-hand circular polarised signal. Even with a transmit antenna gain of 12 dBi the distance that the signal has to travel means that it is very weak when it reaches a receiver on the surface of the Earth. Recovering and decoding the signal will be covered in a little more detail under the discussion of the receiver.

GPS satellite signals use the CDMA (Code Division Multiple Access) spread spectrum technique in which all the satellites transmit their signals in the same frequency band, but each is encoded using a different code (PRN, pseudo-random noise) which allows the receiver to separate them from one another. These codes come from the class of Gold Codes [16] which are pseudo-random binary codes designed to minimise cross-correlation between one another and, therefore, provide the best possible separation between them.

The GPS civil signal is based on a data payload which is transmitted at 50 bit/s (20 ms per bit). This is convolved with the PRN code running at a chip[2] rate of 1.023 MHz. The PRN code used for the C/A signal is truncated to 1023 bits which means that it repeats frequently (every 1 ms) and is therefore easier to use, but truncating it results in a little degradation to performance. The resulting 1.023 MHz data stream modulates the L1 carrier using BPSK (Binary Phase Shift Keying) giving it a bandwidth of just over 2 MHz between the first nulls in the power spectrum.

The GPS military code uses a very much longer PRN code $(6.1871.10^{12}$ bits long) running at 10.23 MHz which is truncated to 7 days' duration. The P code is encrypted to form the Y code (at the same bit rate), hence describing it as the P(Y) code. This is a highly secure code that can only be decoded with knowledge of the key. It is used to BPSK modulate the L2 carrier and an L1 carrier (along with the C/A code). Its carrier is in quadrature to the C/A code carrier (90° out of phase). One of the P(Y) sequences is unmodulated, the other is modulated by the navigation data. The P-code signal occupies a much greater bandwidth: 20.46 MHz between first nulls in the power spectrum.

In the process of modernising GPS a second, civilian, signal has been added in the L2 band. It is only transmitted by the most recent satellites (seven as at July 2012). This additional signal contains two new distinct PRN codes called the Civilian Moderate length (CM) and Civilian Long length (CL) codes. CM is 10 230 bits long and repeats every 20 ms (the data bit interval) and CL is 767 250 bits long repeating every 1.5 s. They are individually transmitted at 511 500 bits/s and multiplexed to yield a 1.023 Mb/s signal. CM is modulated with an upgraded navigation data stream, but CL is unmodulated which allows the receiver to integrate for longer leading to better correlation gain and thus performance.

The future L5 signal is planned as a civilian 'safety of life' signal. It includes two PRN ranging codes:

---

[2] Each bit of the spreading code (PRN) is called a chip.

- The I5 is transmitted in-phase; is 10 230 bits long, is transmitted at 10.23 MHz with a repetition rate of 1 ms and is modulated by a 10-bit Neumann–Hofmann[3] code clocked at 1 kHz.
- The Q5 signal is transmitted in quadrature phase; is 10 230 bits long, is transmitted at 10.23 MHz with a repetition rate of 1 ms and is modulated by a 20-bit Neumann–Hofmann code clocked at 1 kHz.

L5 is transmitted at a higher power level so combined with improved signal structure, wider bandwidth and longer codes should lead to significantly better performance in the future [17].

3.2.4 Receiver operation

Since GPS signals are very weak when they reach the surface of the Earth – with a clear view of the sky and line-of-sight to the satellite, signal power is around −130 dBm – the performance of the receiver is really important. It is important to use a good quality low-noise amplifier in the front end. The tough problem is to acquire the relevant satellite signals, after which continuing to track them is more straightforward.

In order to receive and decode the signal the receiver correlates the received signal with the PRN code for the satellite which transmitted it. This has the effect of restoring the original signal (with data stream) before it was spread by applying the code in the transmitter. It also has the effect of spreading any narrowband interference signals thereby reducing their impact. Signals encoded using a different PRN code are not affected although they remain present as background noise. This operation is the essence of the CDMA technique and leads to an improved signal-to-noise ratio; the amount of improvement is referred to as the coding gain. The coding gain is a function of the code used, in particular its length and the time period over which the correlation is performed. Were it not for this coding gain it would not even be possible to decode the signals from GPS satellites because they are so weak as to lie below the thermal noise floor of the receiver.

---

[3] The Neumann–Hofmann code is a code designed to provide improved signal acquisition, better spreading of interferers, and reduce cross-correlation between different satellite codes.

This illustrates why signal acquisition is the tough problem:

- Each satellite uses a different PRN code and the receiver does not initially know which satellites it is able to receive, and therefore which codes to use.
- In order for the correlation to correctly extract the original signal, not only must the correct code be used, but the codes must be accurately aligned in time – to better than half a chip typically. Since the receiver has no knowledge of the time offset initially it needs to search through all possible offsets. Using half chip steps means 2046 code offsets to search.
- The satellites are travelling through space, the Earth is rotating and the receiver may also be moving, all of which contribute to a Doppler shift in the received signal. Furthermore the receiver oscillator may be a relatively inexpensive device with its own frequency offset. Frequency offsets between received signal and receiver clock lead to correlation errors, and therefore the receiver also has to search across frequency offsets spanning 10 Hz or even more.
- The receiver does not initially have a precise knowledge of time and therefore does not know where data bit boundaries (50 bit/s navigation data stream) lie. Without knowledge of these boundaries correlating across multiple bits is much harder to do, and is less efficient.

In the early days of GPS with less sophisticated receivers the process of searching for and acquiring satellite signals was done sequentially and it took a considerable length of time (several minutes). As electronics has become cheaper and processors more powerful, the acquisition problem has been addressed by using multiple parallel correlators, as well as other assistance techniques, which will be described in more detail later.

However, even if the satellite signals can be acquired fairly quickly, the navigation message (described in the next section) is transmitted at a very low data rate of 50 bit/s, and it takes a significant length of time for the receiver to decode the entire message which is needed in order to compute its position. Speeding up acquisition by providing the navigation message data using a different faster delivery means is the essence of Assisted GPS, described in Section 3.4.

Tracking signals already acquired is a far easier task. It is done by continually monitoring and correcting frequency offset errors to track

Doppler changes as the satellites travel past, and by continually keeping the decorrelating code aligned with the received signal including the data modulation on the signal.

The weak signals and complexity of initial acquisition highlight why it is important for a GPS receiver to have the best possible visibility of the sky and to be shielded from local sources of interference.

3.2.5 Navigation message, almanac and ephemeris

3.2.5.1 General structure

The GPS navigation message comprises 25 frames of 1500 bits, each of which is transmitted in five 300-bit subframes each comprising ten 30-bit words (Figure 3.3). Each 30-bit word consists of 24 data bits and 6 check bits. They are encoded using a 32,26 Hamming code.

Each subframe begins with the TLM (telemetry) word and HOW (handover) word.

The first subframe contains GPS week number, satellite health information and clock correction terms.

The second and third subframes contain ephemeris parameters.

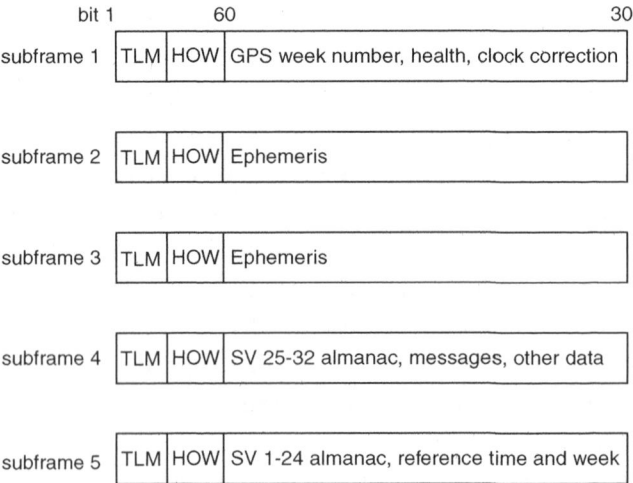

Figure 3.3 Navigation frame structure

The fourth and fifth frames contain almanac data, time correction data, satellite configuration data and additional time reference information. These two subframes consist of 25 pages which are transmitted in turn, one in each frame of the navigation message. The first three subframes are repeated in each frame, with relevant updates to the message content.

Since the data rate is 50 bit/s it takes 6 s to transmit a subframe, 30 s to transmit a frame and 12.5 minutes to transmit the entire navigation message.

3.2.5.2 The TLM (telemetry word)

The TLM comprises a fixed preamble of 8-bits (0x8b) and telemetry data that is only available to authorised users.

3.2.5.3 HOW (handover word)

The handover word (HOW) allows the receiver to 'handover' from C/A code tracking to P(Y) code tracking. The HOW provides time-of-week modulo 6 seconds, antispoofing flag, signal quality flag, and the subframe number. Three bits of the HOW are used to identify which of the five subframes is being transmitted, the frame number can be derived from the time of week (TOW).

3.2.5.4 Week number and TOW (time of week)

A quantity called the Z-count is maintained by the satellites. It comprises a 10-bit week number (the 10 most significant bits), and a time of week (TOW), the least significant 19 bits. This counter counts 'epochs' which occur every 1.5 seconds. The TOW counter runs from 0 to 403 199 then rolls over to 0, representing the epoch count within the week. Week numbers run from 0 to 1023 (10-bit counter) which is a period of nearly 20 years, before rolling over and starting from zero again. The next week number rollover will occur in April 2019.

The TOW transmitted in the HOW is the actual TOW modulo 4 (17 bits) since the HOW is repeated in each subframe, which is every 6 seconds. The TOW transmitted is the TOW representing the TOW at which the next subframe starts, and is obtained simply from multiplying the transmitted 17-bit TOW by 4.

3.2.5.5 Subframe 1: clock correction data

This subframe contains the week number and clock correction data in the form of polynomial coefficients describing how the satellite clock varies with time. It also includes the clock reference time, which is the time origin used to compute the clock error; the ionospheric group delay; and the issue number of the clock data.

3.2.5.6 Subframes 2 and 3: ephemeris

Ephemeris is the data required to compute the exact position and velocity of the satellite required for the navigation solution. This information is very precise and is typically only useful for a period of about four hours, and only applies to the satellite transmitting it. The components of the ephemeris data are listed below:

Term	Description	Units
M_0	mean anomaly at reference time	semicircle
Δn	mean motion correction	semicircle/s
e	eccentricity	dimensionless
$\sqrt{a}$	square root of semi-major axis	$m^{\frac{1}{2}}$
Ω_0	longitude of ascending node of orbit plane at weekly epoch	semicircle
i_0	inclination angle at reference time	semicircle
ω	argument of perigee	semicircle
$d\Omega/dt$	rate of right ascension	semicircle/s
di/dt	rate of inclination angle	semicircle/s
C_{uc}	cosine harmonic correction term of latitude	rad
C_{us}	sine harmonic correction term of latitude	rad
C_{rc}	cosine harmonic correction term of orbit radius	m
C_{rs}	sine harmonic correction term of orbit radius	m
C_{ic}	cosine harmonic correction term of angle of inclination	rad
C_{is}	sine harmonic correction term of angle of inclination	rad
t_{0e}	reference time	s
$IODE$	issue number of ephemeris data	dimensionless

3.2.5.7 Subframe 4: almanac and special messages

Almanac for satellites 25 to 32 as well as special messages, ionospheric correction terms and coefficients to convert GPS time to UTC. Almanac data is similar to ephemeris except that it is less precise and remains usable for much longer periods of time – potentially months, and information for all satellites is transmitted.

3.2.5.8 Subframe 5: almanac

Almanac data for satellites 1 to 24.

3.2.6 Location computation

The first step to determining the receiver location is to calculate the positions of the satellites, and then using pseudo-range measurements the receiver location and time offset can be calculated.

3.2.6.1 Satellite positions

The ephemeris parameters are used to calculate the satellite position. Solving this is known as the Kepler problem after Johannes Kepler who published his first two laws of planetary motion in 1609 and his third in 1619:

1. The orbit of every planet is an ellipse with the Sun at one of the two foci.
2. A line joining a planet and the Sun sweeps out equal areas during equal intervals of time.
3. The square of the orbital period of a planet is directly proportional to the cube of the semi-major axis of its orbit.

These equations are only strictly true for a weightless body (e.g. planet) orbiting another body (e.g. Sun) in the absence of any other perturbing forces, such as other planets and objects. In the case of satellites orbiting the Earth, factors such as Moon and Sun have a significant effect on the precise orbits of the satellites. Computing the precise orbits requires knowledge of the orbits and positions of any such perturbing bodies. Furthermore there are other forces acting on the satellites including solar radiation pressure and internal factors such as release of internal gas from the satellite, tidal variations on Earth and others. However, it has been shown that all of these external forces and accelerations can be collected together and defined in terms of the two-body solution.

For the standard two-body equations it can be shown that there are six constants of integration. When the more complex multi-body system of a satellite orbiting the Earth is modelled as a two-body solution, these six constants of integration exist, but they vary with time. This is the approach taken in the GPS ephemeris message: the six constants of integration are provided along with a reference time at which they are applicable as well as a characterisation of how they change over time. From these the GPS receiver can compute the orbital characteristics and hence the satellite position. This time-changing nature of the parameters used to model the satellite orbit explains why the ephemeris data has a limited useful life.

It is beyond the scope of this book to go into the details of Kepler's equations, or precisely how the GPS receiver solves them in order to compute the satellite position. To do so requires iterative numerical techniques and the reader is referred to some of the excellent texts on the subject, such as Kaplan [18].

As a summary the equations used to compute the satellite position, in ECEF coordinates, are presented below:

WGS84: universal gravitational parameter m^3/s^2

$$\mu = 3.986005.10^{14} \tag{3.1}$$

WGS84: Earth rotation rate, rad/s $\quad \dfrac{d\Omega_e}{dt} = 7.292115167 \cdot 10^{-5} \tag{3.2}$

Semi-major axis $\quad a = \left(\sqrt{a}\right)^2 \tag{3.3}$

Corrected mean motion, rad/s $\quad n = \sqrt{\dfrac{\mu}{a^3}} + \Delta n \tag{3.4}$

Time from ephemeris reference epoch $\quad t_k = t - t_{0e} \tag{3.5}$

Mean anomaly $\quad M_k = M_0 + nt_k \tag{3.6}$

Kepler's equation for eccentric anomaly $\quad M_k = E_k - e\sin E_k \tag{3.7}$

True anomaly from sine $\quad \sin v_k = \dfrac{\sqrt{1 - e^2}\, \sin E_k}{1 - e \cos E_k}$ $\hspace{2cm}$ (3.8)

True anomaly from cosine $\quad \cos v_k = \dfrac{\cos E_k - e}{1 - e \cos E_k}$ $\hspace{2cm}$ (3.9)

Eccentric anomaly from cosine $\quad \cos E_k = \dfrac{e + \cos v_k}{1 + e \cos v_k}$ $\hspace{1.5cm}$ (3.10)

Argument of latitude $\quad \phi_k = v_k + \omega$ $\hspace{3cm}$ (3.11)

Second harmonic correction to latitude

$\qquad \delta\phi_k = C_{uc} \cos 2\phi_k + C_{us} \sin 2\phi_k$ $\hspace{2.5cm}$ (3.12)

Second harmonic correction to radius

$\qquad \delta r_k = C_{rs} \sin 2\phi_k + C_{rc} \cos 2\phi_k$ $\hspace{2.5cm}$ (3.13)

Second harmonic correction to inclination

$\qquad \delta i_k = C_{is} \sin 2\phi_k + C_{ic} \cos 2\phi_k$ $\hspace{2.5cm}$ (3.14)

Corrected argument of latitude $\quad \mu_k = \phi_k + \delta\phi_k$ $\hspace{2cm}$ (3.15)

Corrected radius $\quad r_k = a(1 - e \cos E_k) + \delta r_k$ $\hspace{2cm}$ (3.16)

Corrected inclination $\quad i_k = i_0 + \left(\dfrac{di}{dt}\right) t_k + \delta i_k$ $\hspace{1.8cm}$ (3.17)

X coordinate in orbit plane $\quad x_p = r_k \cos \mu_k$ $\hspace{2.3cm}$ (3.18)

Y coordinate in orbit plane $\quad y_p = r_k \sin \mu_k$ $\hspace{2.3cm}$ (3.19)

Corrected longitude of ascending node

$\qquad \Omega_k = \Omega_0 + \left(\dfrac{d\Omega}{dt} - \dfrac{d\Omega_e}{dt}\right) t_k - \dfrac{d\Omega_e}{dt} t_{0e}$ $\hspace{2cm}$ (3.20)

ECEF X coordinate $\quad x_k = x_p \cos \Omega_k - y_p \cos i_k \sin \Omega_k$ $\hspace{1.5cm}$ (3.21)

ECEF Y coordinate $\quad y_k = x_p \sin \Omega_k + y_p \cos i_k \cos \Omega_k$ $\hfill$ (3.22)

ECEF Z coordinate $\quad z_k = y_p \sin i_k$ $\hfill$ (3.23)

3.2.6.2 Calculating receiver position

As described earlier GPS signals are spread spectrum signals created by multiplying the data signal by a PRN code which has the characteristic of spreading the signal across a wider bandwidth, and since each satellite uses a different code the signals from the respective satellites can be recovered in the receiver by correlating the received signal with the codes for each satellite – provided that proper code alignment and frequency offset can be found (the acquisition problem).

Assuming that proper code alignment has been achieved this allows the receiver to determine the precise time of arrival (according to its local clock) of the satellite signal. This is achieved by:

1. Using the GPS time from the data message to compute the bit time and from that the code time (remember that the PRN code is 1023 bits long and repeats every 1 ms);
2. Measuring the bit count and fractional bit offsets within the PRN code, corresponding with the correlation peak in the receiver.

To this measured time of arrival of the satellite signal is added any corrections (satellite clock corrections) and errors (ionospheric delay etc.) reported by the satellite. Since the time of transmission of the signal by the satellite as measured by the satellite clock is known to the receiver – the satellite transmits the signals aligned with its own clock – this allows the receiver to compute pseudo-ranges for each measured satellite.

Note that it is important to measure the time of arrival of the signal as accurately as possible in order that measurement errors do not degrade the position calculation: the signal travels at the speed of light, approximately 3 ns for every metre. Put another way, each chip of the PRN code represents approximately 300 metres in range.

So now having computed the satellite positions, a set of equations representing the pseudo-ranges for all visible satellites can be constructed as follows:

$$\rho_i = \sqrt{(x_i - x_u)^2 + (y_i - y_u)^2 + (z_i - z_u)^2} + ct_u \qquad (3.24)$$

where (x_i, y_i, z_i) is the satellite position, (x_u, y_u, z_u) is the receiver position and t_u is the clock offset of the receiver relative to GPS time.

As can be seen this is a non-linear equation with four unknowns (x_u, y_u, z_u, t_u), and therefore to solve it one needs at least four sets of equations, in practice measurements of at least four satellites. There are several different ways to go about solving this set of equations:

1. closed-form solutions;
2. iterative techniques based on a linearised form of the equations;
3. iterative techniques using direct minimisation algorithms;
4. Kalman filtering.

The last of these is commonly used in combination with velocity estimates as a way of improving performance by using a time series of measurements.

Note that the position returned by solving this set of equations is the receiver position (more precisely the position of the phase centre of the receiver antenna) in ECEF coordinates. In order to deliver a latitude, longitude and height the position needs to be converted to WGS84 coordinates based on the reference ellipsoid as described in Chapter 2.

It is beyond the scope of this book to describe the details involved in constructing and solving the positioning equations, and the user is referred to the more specialised texts on the subject.

3.2.6.3 Calculating receiver velocity

In some receivers the velocity of the receiver is estimated using a simple difference equation between successive positions. This approach can be satisfactory provided the velocity is essentially constant over the difference interval, and provided successive positions are spaced apart by significantly further than the errors in the positions. For this method we can write:

$$\dot{u} = \frac{du}{dt} = \frac{u(t_2) - u(t_1)}{t_2 - t_1} \tag{3.25}$$

where u represents the position of the receiver and the derivative of u is its velocity.

However, there is another way of computing velocity, and that is done using the Doppler measurements of the satellite signals. Doppler frequency is a function of the speed at which a receiver is travelling towards (or away from) a transmitter. Since the satellites are moving too, the measured Doppler frequencies are a function of the combined speed of the receiver and satellite in a direction along the vector connecting the two.

So we have a combined velocity given by:

$$\mathbf{v}r = \mathbf{v} - \dot{\mathbf{u}} \tag{3.26}$$

which, if we form the dot product with a unit vector pointing along the line of sight from the receiver to the satellite gives us the radial velocity in this direction, which we can equate to the frequency difference as follows:

$$\frac{\Delta f}{f_T} c = (\mathbf{v} - \dot{\mathbf{u}}) \cdot \mathbf{a}. \tag{3.27}$$

Since the frequency difference is formed from the sum of the Doppler frequency and the frequency error between the satellite clock and the receiver clock, we can rearrange this equation to arrive at a system of equations in which, for the ith satellite, we have:

$$0 = \frac{f_{Ti}}{(f_{Ri} - f_{Ti})} \frac{1}{c} (\mathbf{v} - \dot{\mathbf{u}}) \cdot \mathbf{a}. \tag{3.28}$$

The value of f_{Ti} is the operating frequency of the signal plus any clock corrections from the ephemeris data. The satellite velocity is obtained from the ephemeris data applying appropriate orbital equations. The unit vector is obtained assuming a position is known (due to the distance between receiver and satellite, an approximate position is usually good enough). This leaves us with four unknowns for the velocity of the receiver and the clock offset f_{Ri} of the receiver. Using measurements from four or more satellites a system of four or more equations (one for each satellite) can be constructed and solved to give the receiver velocity.

The detailed method of constructing and solving these equations is beyond the scope of this book and the reader is referred to one or more of the specialist texts on the subject.

3.2.7 Time management and clocks

In the position calculation shown earlier we solved for the unknown position and unknown receiver clock offset. Having solved for this we have an accurate local representation of GPS time in the receiver.

This ability to accurately distribute time to GPS receivers is extremely useful for many time synchronisation tasks such as synchronisation of mobile network base stations, or for accurate time synchronisation in computers for applications requiring very precise time stamps – such as in financial transactions. However, it is worth being aware of exactly what GPS time is and how it is managed.

GPS time is referenced to UTC (coordinated universal time) and was coincident at midnight at the start of 6 January 1980. However, GPS time is a continuous scale and it is, therefore, not corrected for leap seconds. Therefore over time it has moved apart from UTC, although the difference is kept modulo 1 second. At the time of writing GPS time was ahead of UTC by 15 seconds.

GPS time is managed to be within 1 μs of UTC modulo 1 second, although in practice it is generally better than 50 ns. The polynomial time correction factors provided in the navigation data message can be used to provide a good estimate of this error, and, therefore, knowing the whole second offset between GPS time and UTC it is possible for a receiver to provide a very good estimate of UTC to the user. The accuracy of this time is largely determined by the performance of the receiver in its operating environment. A good time transfer receiver in a static position with a clear view of the sky can provide better than 30 ns accuracy.

3.3 High precision GPS systems

3.3.1 Differential GPS

Differential GPS (DGPS) is a technique for reducing the error in the GPS position by using additional measurements made by a reference receiver at a known position. The reference receiver measures the GPS satellite signals and since its location is known it is able to calculate corrections for satellite ephemeris and clock errors, and ionospheric and tropospheric delay errors. However, it is unable to correct many other sources of errors such as multipath and receiver errors for the mobile receiver.

The corrections calculated by the reference receiver are sent to the mobile receiver using a suitable communications channel and it applies them to its own measurements. Provided that the two receivers are in similar environments (within a couple of hundred kilometres usually qualifies) and they can receive the same satellite signals, the corrections help reduce the error and lead to an improved position estimate for the mobile receiver.

Differential GPS was hugely beneficial in the past when selective availability (SA) was turned on. SA was a technique used to add noise deliberately to the GPS signals and thereby degrade the accuracy to around 100 metres. In May 2001 SA was turned off and it has not been used since. Differential GPS proved to be capable of removing almost all the error caused by SA, but it leads to only a modest improvement over GPS without SA.

Differential corrections can be provided by a local reference receiver, but there are a number of systems offering wide area coverage through the systematic provision of correction information. These are often lumped together under the term SBAS (Space-Based Augmentation Systems). The best known include WAAS (Wide-Area Augmentation System) and EGNOS (European Global Navigation Overlay System), but there are a number of others covering other regions. They tend to have a regional focus as illustrated in Figure 3.4.

Correction data supplied by these SBASs are usually provided via a geostationary satellite covering the region and linked to a network of ground monitoring stations used to measure the satellite errors. A suitably equipped GPS receiver is able to receive the transmitted corrections from the geostationary satellite and apply them to its own measurements. A key aspect of these systems is for integrity monitoring, so apart from improved positioning accuracy, they are an important element in monitoring and checking for signal integrity, especially for mission critical systems in which confidence in the GPS position is essential.

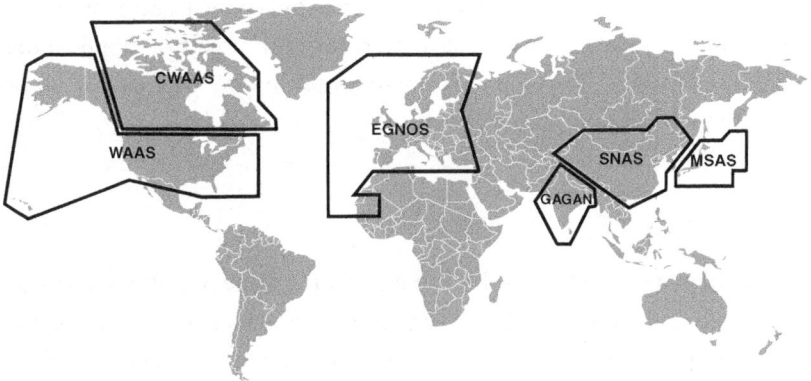

Figure 3.4 Selected GNSS Augmentation Systems

3.3.2 Carrier phase tracking

Another way of obtaining much higher precision is to measure the carrier phases of the L1 and L2 signals and use them to compute a high accuracy position. Phase measurements can give very good precision, but are susceptible to multipath, and, most significantly, have a 'cyclic ambiguity' factor that needs to be resolved. A more detailed description of phase measuring systems is included in Chapter 4 on radiolocation techniques.

The 'cyclic ambiguity' arises because the phase measurement is a modulo-2π (radians) measurement over the wavelength of the signal being measured. In the case of GPS, L1 has a wavelength of approximately 19 cm and L2 has a wavelength of approximately 24.5 cm. These mean that carrier phase measurements at the centimetre precision can readily be made, but in order to be useful for positioning the cyclic ambiguity needs to be resolved by localising the receiver to within about 20 cm before the phase measurement becomes useful. One technique that can be used to help is to compute and measure the signal formed as the difference between the L1 and L2 carriers. This corresponds to a wavelength of approximately 86 cm.

By adding additional redundant satellite measurements, and searching over time as the satellites move it may be possible to resolve these ambiguities. However, real situations in which the receiver is moving and in a 'normal' environment make carrier phase measurements rarely useful for single receiver use. However, for high precision applications such as with RTK (Real-Time Kinematic) for survey use, carrier phase measurements play an important role.

3.3.3 RTK GPS

RTK (real-time kinematic) GPS is based on the same principles as differential GPS, except that in this case the fixed reference receiver measures the carrier phases and broadcasts them to the mobile receivers. In an RTK system the reference receiver position needs to be known very precisely (centimetres), and the positions of the moving receivers can then be computed to similar levels of accuracy.

Broadcasting the carrier phase information generally needs a dedicated radio telemetry link between fixed and moving receivers, and it involves sending a significant amount of data over this link. The fixed and mobile receivers are typically quite close – hundreds of metres, rather than hundreds of kilometres apart – and they need to be able to receive and measure the same satellites. Multipath badly affects phase measurements and can seriously degrade the performance of an RTK system, so they are generally only really useful in open outdoor conditions.

3.4 Assisted GPS

3.4.1 Introduction

Assisted GPS (A-GPS) capability is used for two common reasons:

1. To allow the GPS receiver to find the satellites and complete a fix under difficult reception conditions such as indoors; and/or
2. To reduce the time-to-first-fix (TTFF) of the receiver, thereby allowing it to compute a position far quicker than normal from either cold or warm start.

A-GPS is not a means for getting higher accuracy.

It arose as a mechanism to assist GPS receivers that are only required to do one-off or occasional position fixes, such as in mobile consumer devices, such as cellular phones. If left on the GPS receiver has a significant impact on the battery life of the device. Therefore the GPS receiver is powered down until a position fix is required.

In order to support A-GPS three additional system components are required:

1. a GPS reference receiver that is able to capture ephemeris data and other GPS data for all satellites in view;
2. a server with which the mobile receiver communicates in order to retrieve satellite data relevant to its location (approximately) when it needs to make a fix;
3. a communications network, such as via a cellular data bearer which it is able to use to fetch the assistance data.

The first two are generally provided by a service provider and access to the assistance data is charged for according to an agreed tariff.

3.4.2 Satellite information to assist acquisition and TTFF

Receiving the satellite ephemeris information from an assistance server saves the receiver from having to decode the received satellite signals and extract the data – which at 50 bit/s takes quite a long time. It also allows the receiver to use a much narrower frequency search window since the satellite position and Doppler offset are known and only the receiver's own clock offset and Doppler are unknown. Furthermore it may allow the receiver to do coherent integration over a much longer time period, provided it knows the bit sequence used to modulate the satellite signal, and therefore it is possible to acquire weaker satellite signals than is normally possible.

3.4.3 Using cellular network signals to improve TTFF

Cellular networks generally operate to quite close frequency tolerances (typically ± 0.05 ppm), and in some cases time synchronisation is also required for base stations.

A cellular mobile device can take advantage of this, since it is able to frequency align its clock with the local base station clock. Having a local receiver clock accurate to 0.05 ppm reduces the frequency range over which to search for satellite signals. However, the search across the code space time offset is still required – unless absolute time synchronisation to better than 1 ms can be achieved.

Techniques for very precise time transfer to the mobile device, using the capabilities of the cellular network have been proposed and shown to be feasible. Using these techniques it is possible to reduce the initial code search space to just a few chips. Cellular positioning techniques based on observed time-of-arrival techniques are usually able to achieve this level of absolute time synchronisation.

3.5 Brief descriptions of other GNSSs

3.5.1 GLONASS

GLONASS is the system operated by the Russian government. It had fallen into a state of disrepair, but in the last ten years has been brought back to full operational status again, with all 24 satellites declared operational in December 2011 giving it global coverage again.

As for GPS it comprises space, ground and user segments. The space segment comprises 24 satellites arranged in three orbital planes inclined at 64.8°, the satellites being evenly distributed in each plane. They orbit at a height of approximately 19 100 km, and have an orbital time of approximately 11 hours 15 minutes and 44 seconds. The increased inclination compared with GPS gives better coverage in very high latitudes.

The radio interface is different from GPS in that it uses an FDMA (frequency division multiple access) scheme rather than CDMA. However, the frequencies used are in the same bands as GPS: L1 (1598–1606 MHz) and L2 (1243–1249 MHz).

Similar to GPS, GLONASS uses two codes: a coarse code (0.511 MHz) and a precision code (5.11 MHz). Each satellite transmits the same code, but on a different frequency. The frequencies used are given by:

$$f_{K1} = f_{01} + K\Delta f_1$$
$$f_{K2} = f_{02} + K\Delta f_2$$
$$f_{01} = 1602\,\text{MHz}, \Delta f_1 = 562.5\,\text{kHz}$$
$$f_{02} = 1246\,\text{MHz}, \Delta f_2 = 437.5\,\text{kHz}.$$

(3.29)

K takes values from -7 to $+6$, with 5 and 6 reserved for technical purposes. The same frequencies are reused on opposite sides of the orbits, hence the maximum constellation of 24 satellites.

Received power levels on the surface of the Earth are very similar to GPS, and the accuracy obtained from GLONASS is about 50% less than GPS. The terrestrial reference frame used by GLONASS is PZ 90 rather than the more familiar WGS84 used by GPS.

3.5.2 Galileo

Galileo is the third GNSS largely conceived and funded by the countries of Europe and the European Space Agency. Knowing some of the limitations of GPS has helped to improve the system, although many of the improvements adopted by Galileo are being adopted back into GPS through its modernisation phases.

However, one of the major differences with Galileo is that it was conceived to be a commercial system and therefore it has been designed around four main services:

- The Open Service (OS) carrying open signals that are free to use, mainly for mass market applications.
- The Safety of Life service (SoL) which provides integrity data in addition to the OS. It is aimed at transport and similar applications where lives could be at stake.
- The Commercial Service (CS) which has two additional signals and faster data rates giving potentially better accuracy. This is a subscription service.
- The Search and Rescue service (SAR) will support the international SARSAT system for humanitarian search and rescue.

A Galileo constellation will comprise 30 satellites (of which 3 are spares) in MEO (medium Earth orbit) at approximately 23 600 km altitude with an orbital period of about 14 hours. There are three orbital planes each having 10 satellites. The orbital inclination is 56°, slightly more than GPS, which with the extra altitude gives slightly better polar and high latitude coverage.

System specifications were agreed after intense negotiations with stake-holders around the world and the first satellite was launched in 2005. Since then the project has been delayed and has been subject to much controversy but in 2010 agreement to proceed was reached and on current plans satellite launches should commence in 2012 for a fully operational date of 2018.

Galileo signals will operate in three bands centred on 1.191 GHz, 1.278 GHz and 1.575 GHz. Two of them are shared with GPS and the signals are designed to interoperate with GPS. Within these bands are 11 signals occupying a total of more than 120 MHz combined channel bandwidth. However, there are still negotiations going on with China in relation to the Compass system which also uses some of the same bands.

One of the important features of the Galileo signals is that they include unmodulated pilot tones. These allow much longer integration times without needing to decode the modulation data for the navigation message. This allows for better signal acquisition, particularly in challenging conditions with weak signals.

Code rates are the same as for GPS, and each satellite is modulated with its own unique code which is also the means for identifying the satellite.

3.5.3 Compass

Compass, also known as Beidou-2, is a Chinese GNSS similar to GPS and Galileo. Initially China was part of the Galileo programme, but politics forced the parties to split over security related fears, so China has embarked on its own independent system.

The Compass navigation system will comprise a total of 37 satellites: 5 geostationary and 32 non-geostationary, providing complete coverage of the globe. On 27 December 2011 the system was declared operational with 14 satellites providing navigation capability in the Asian region. The

14 operational satellites comprise five geostationary satellites positioned at 58.75°E, 80.00°E, 110.50°E, 140.00°E and 160.00°E. Four of the non-geostationary satellites are Medium Earth Orbit (MEO) and the other five are Inclined Geosynchronous Satellite Orbit (IGSO).

The system uses the China Geodetic Coordinate System 2000 (CGCS2000). It is aligned with the IERS pole and reference meridians and has the following ellipsoid parameters:

Semi-major axis: $a = 6\,378\,137.0$ m
Gravitational constant: $GM = 398\,600.4418 \cdot 10^9$ m^3/s^2
Flattening: $f = 1/298.257222101$
Earth rotation rate: $\omega = 7.2921150 \cdot 10^{-5}$ rad/s

Its time system uses SI seconds and a zero point at UTC 1 January 2006, 00:00. It is synchronised to UTC to an accuracy of 100 ns modulo 1 s.

The signals use CDMA with complex structures similar to Galileo and Modernised-GPS. It will also offer two levels of positioning: coarse being free-to-air and freely available; and precise for restricted (military) use. From 27 December 2011 the basic B1 signal, having the following characteristics, is available for navigation:

Carrier frequency	1561.098 MHz
Modulation	QPSK
Bandwidth	8 MHz (3 dB)
Signal strength at ground	> -133 dBm for channel I, above 5° elevation
Ranging code rate	channel I: 2.046 Mcps
Ranging code length	channel I: 2046 chips
Data rate GEO satellites	channel I: 500 bps
Data rate MEO/IGSO satellites	channel I: 50 bps with secondary code rate 1 kbps
Access mode	CDMA
Polarisation	right-handed circular polarisation

It is expected to be 2020 before the system is fully deployed and operational, but interested parties are invited to participate with testing and an interface

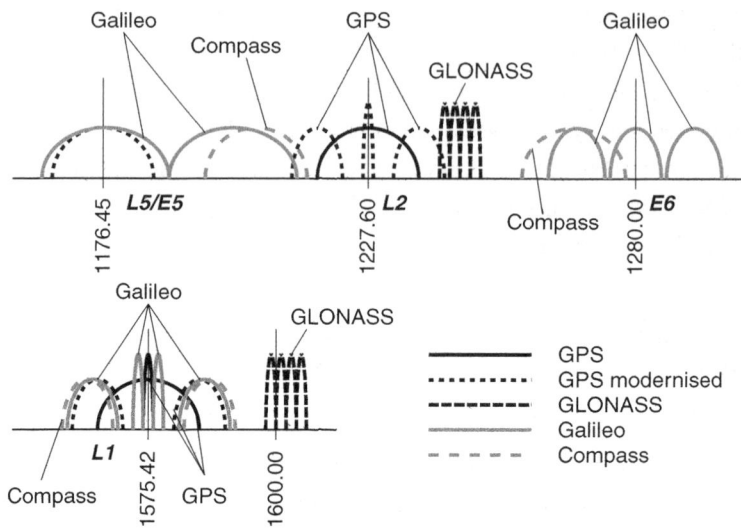

Figure 3.5 Current and proposed frequencies for GNSS

document [19] has been released for this purpose. A more detailed description of Compass can be found in reference [20].

3.5.4 Regional satellite navigation systems

In addition to the four GNSSs described above, there are several regional systems:

- Beidou-1, China, the forerunner of Compass described above;
- DORIS, Doppler Orbitography and Radio-positioning Integrated by Satellite, France;
- IRNSS, The Indian Regional Navigational Satellite System;
- QZSS, The Quasi-Zenith Satellite System, Japan.

Further discussion of these systems is beyond the scope of this book.

3.5.5 Summary of GNSS spectrum allocations

Figure 3.5 is a diagrammatic summary of frequency bands allocated to and used by the four main GNSSs described above.

4   Radiolocation technologies

There are several techniques for using radio signals to determine the position of an object:

1. Angle of arrival of a received signal indicates the direction to the transmitter. On its own this does not allow position to be determined as one also needs to know the distance between transmitter and receiver.
2. Direct inference of range. Range can be inferred in a number of ways: using signal strength; using techniques for measuring the time-of-flight of the radio signal; using phase of the received signal as a measure of range (assuming techniques for synchronising transmit and receive clocks can be implemented).
3. Measuring the time of arrival of a radio signal. In the absence of a definitive time reference (usually the case) it is necessary to measure arrival times of two or more signals and then by comparing them a position can be computed. Time Difference of Arrival or Observed Time Difference of Arrival is the most common way of resolving clock uncertainties.

This chapter will look at each of the main techniques in turn, identifying their strengths, weaknesses, advantages and disadvantages, with an emphasis on their practical use in real-world positioning systems.

4.1 Angle of arrival

Measuring the angle of arrival (AOA) of a radio signal is one of the oldest methods of locating a source and for many applications is still the preferred method. Radar systems are based on the principle of angle or bearing to target; they make use of a highly directional antenna sweeping the area and looking for echoes from reflective objects in the scanned area. Homing or beacon finding devices, often used for tracking wild animals or for stolen car recovery, may use angle of arrival to locate the source, a radio transmitter attached to the object being tracked, and then follow the direction of the radio signal until the quarry has been found. Most optical systems for tracking a target, whether with moving or fixed cameras are based on measuring the angle to the target.

4.1.1 Principle of angle of arrival

Given a single measurement of bearing and range the position of the target object can be determined provided that the position and orientation of the sensing device are known. See Figure 4.1.

The position of the target is computed as follows:

$$
\begin{aligned}
x_r &= x + r\cos\theta \\
y_r &= y + r\sin\theta
\end{aligned}
\tag{4.1}
$$

If angles to the target can be measured from two or more geographically distinct positions the position of the target can be determined without knowing the ranges (Figure 4.2).

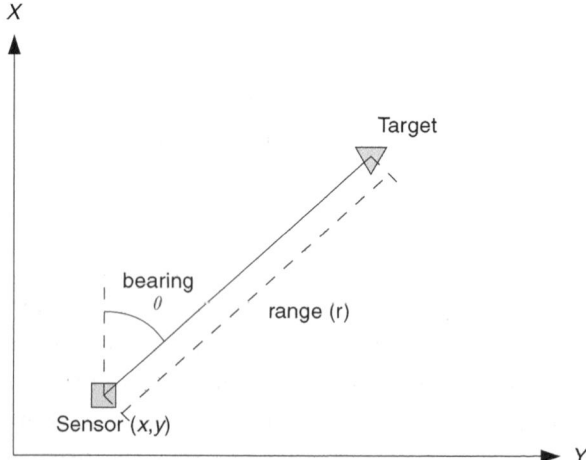

Figure 4.1 Position using angle and range

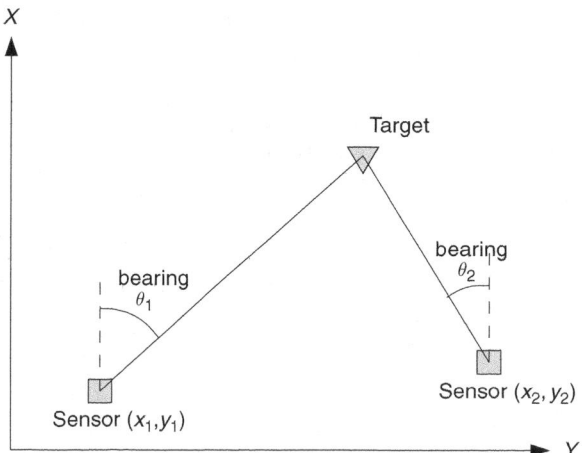

Figure 4.2 Angle measurements from two points

The position of the target is given by the solution to the following pair of simultaneous equations:

$$\tan \theta_1 = \frac{y_T - y_1}{x_T - x_1}$$
$$\tan \theta_2 = \frac{y_T - y_2}{x_T - x_2}. \tag{4.2}$$

4.1.2 Measuring the angle of arrival

The two most usual methods for measuring the angle of arrival of a radio signal are:

1. use a directional antenna and rotate it seeking maximum signal strength (Figure 4.3);
2. use an antenna with steerable beam to search for the direction of maximum signal strength.

Other methods, including the use of interferometry, both with multiple antennas and distributed sensor devices, for example Amundson [21], continue to be the subject of further research because of the complexity of implementing sophisticated antennas in small low-cost sensor devices, and also to reduce the dependence on knowing the precise orientation of the sensor.

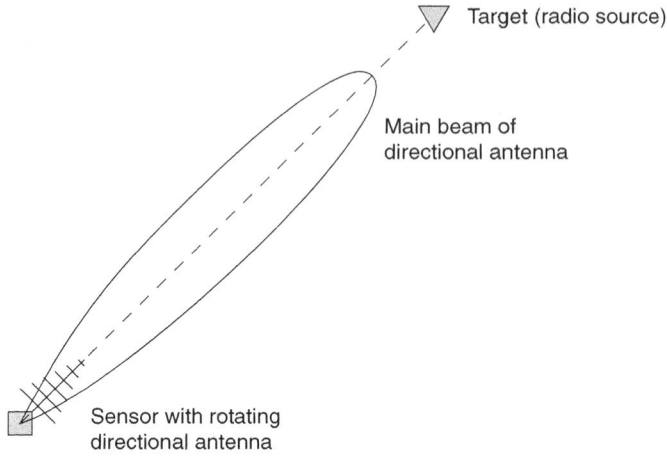

Figure 4.3 Narrow beam antenna used for AOA

4.1.3 Advanced techniques for AOA

A number of techniques for joint estimation of time of arrival (TOA) and direction of arrival (DOA) are being developed. These include ESPRIT (estimation of signal parameters via rotational invariance technique), JADE (joint angle and delay estimation) and SAGE (space-alternating generalised expectation maximisation). Li [22] provides a study of the comparative performance of some of these algorithms.

4.2 Received signal strength (RSSI) to determine distance

By far the most widely used measurement for indoor positioning using wireless networks – Wi-Fi, ZigBee etc. – is RSSI. However, reflections, scattering, variable attenuation through objects, attenuation of the human body, non-isotropic antenna radiation patterns and many other factors result in large variations in the measured RSSI. Furthermore typical low-cost wireless radios are unable to measure RSSI more accurately than about ±3 dB. So while RSSI can be used as a method for positioning wireless devices in a network when these variables can be controlled, this is not true for most real-world applications using small low-power body-worn devices with non-ideal antennas. For these applications RSSI is a poor indicator of range.

Section 6.6 briefly discusses the use of 'fingerprinting' techniques which attempt to measure and map signal strength across the entire region of interest, so that measurements can be associated with a particular location using pattern matching techniques. This highlights some of the complexities in using RSSI for real-world location applications.

Nevertheless, RSSI can and does have a role to play in positioning systems. The basis of RSSI measurements is that under open unobstructed conditions, transmitted signals become weaker as they propagate away from the transmitting antenna. There are a number of different path loss models used for different kinds of signals and environments, but they tend to follow the general formula for *RSSI* at distance *d* relative to *d0*:

$$RSSI(d) = P_t - P_{d0} - 10\eta\log_{10}\frac{d}{d0} + X_\sigma \qquad (4.3)$$

where P_t is the transmitted power, P_{d0} is the power at *d0*, η is the path loss exponent and X_σ is Gaussian random noise with variance σ^2. The path loss exponent is chosen to model the radio channel and environment. The units of RSSI and power are usually expressed in dBm (dB relative to 1 mW).

What this equation tells us is that the error in estimated range given a measurement margin in dBm for RSSI is much smaller for smaller ranges and it grows rapidly as range increases. For example at a range of 4 metres, RSSI might indicate that the range is between 2 m and 8 m, but for a range of 100 metres it would indicate that it is between 50 m and 200 m. Therefore RSSI is a much better position indicator for near distances. The actual error margins depend on the particular parameters of the path loss model for the environment, and are best determined empirically using the actual radio equipment used by the sensors.

4.3 Time-of-flight range measurement

Time-of-Flight is a technique used to measure the distance separating a pair of devices. Each device has both transmitter and receiver, and the

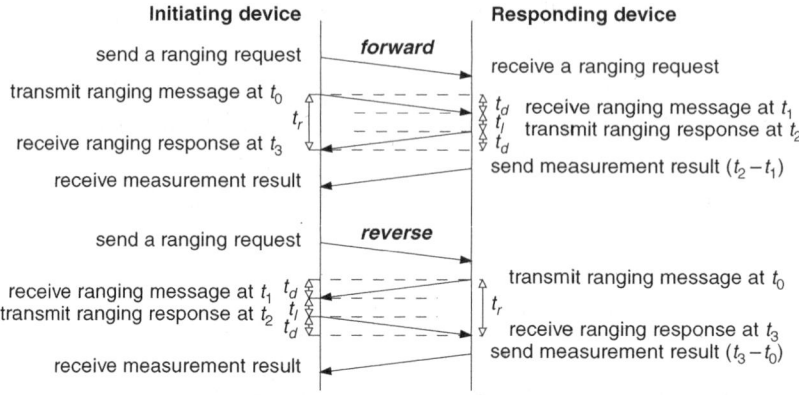

Figure 4.4 Signalling for time-of-flight measurement

process of measuring time-of-flight involves an exchange of signals and measurement results between the two cooperating devices. The basic time-of-flight measurement procedure is shown in Figure 4.4.

The principle is that an originating device initiates a ToF measurement with a request to a responding device. A pair of range messages is exchanged and the responding device sends its measurements – delay from receiving to responding in the case of a forward sequence or the time from sending to receiving in the case of a reverse sequence. Using these measurements the originating device computes the range as follows:

$$\begin{aligned}
t_r &= t_3 - t_0 \\
t_l &= t_2 - t_1 \\
t_d &= \frac{t_r - t_l}{2} \\
d &= ct_d
\end{aligned} \tag{4.4}$$

where c is the speed of light (the propagation speed of radio signals in free air). From the Equations (4.4) it can be seen that the two times that are subtracted are measured by the originator and responder respectively, each using only their own local clock. Therefore, there is no need for the two clocks to be synchronised and each device can use its own free-running clock.

For best accuracy both forward and reverse measurements are made and the average of the two is used as the measured range. The reason for doing this is to compensate for clock offsets between originator and responder. For example, the propagation time between two devices 10 m apart is approximately 33 ns giving a double trip time of 66 ns, and with message times of 'one or two milliseconds' plus response time we might have a total response time t_r of around 4 ms. Therefore the message time plus response latency dominates the measurement. Assuming a clock stability of ± 10 ppm we could see as much as 80 ns of error if both clocks were at extreme tolerance offset. Admittedly this is unlikely, but errors of several tens of nanoseconds are common, and these lead to errors of several metres.

In order to improve performance further it is also common to make a number of round-trip time-of-flight measurements in a single session and to average them. This also helps to make the whole exchange more efficient since only a single initiation and (usually) only a single results response message is required and only the actual round-trip measurements are repeated.

Time-of-flight is an effective method for making point-to-point range measurements between pairs of devices and as such it can be used in many applications provided there is only a small number of devices in the network and that high positioning update rates are not required. It is unsuitable for larger networks or networks requiring high position update rates.

4.4 Time of arrival

Suppose we have three devices, A, B and C as shown in Figure 4.5 with B and C at known positions and the position of A unknown. Suppose that their clocks are synchronised so that all time measurements made by them can be related without the need to account for clock offsets. At time t_a device A sends a signal to both B and C which measure the times at which it arrives as t_b and t_c.

Subtracting the transmit time from the receiving time and multiplying by the speed of propagation gives the two ranges r_b and r_c and from this

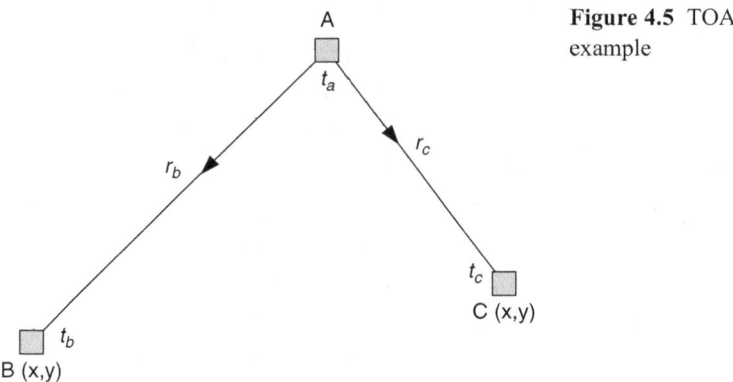

Figure 4.5 TOA example

we can construct a pair of simultaneous equations that allow the unknown position (x_a, y_a) of A to be solved:

$$\begin{aligned} r_b = c(t_b - t_a) &= \sqrt{(x_a - x_b)^2 + (y_a - y_b)^2} \\ r_c = c(t_c - t_a) &= \sqrt{(x_a - x_c)^2 + (y_a - y_c)^2} \end{aligned}. \qquad (4.5)$$

Of course the real challenge in systems strictly based on TOA is clock synchronisation. One approach is to use different mechanisms for synchronisation and range measurement. An example of such a system is one in which radio signals are used for clock synchronisation and audio signals are used to measure ranges is presented by Sallai in [23].

4.5 Time difference of arrival

4.5.1 Basic TDOA system

Time difference of arrival (TDOA) or Observed time difference of arrival (OTDOA) is a method of eliminating, or rather estimating, the clock offset in the mobile device in the course of solving for its position. To do so we need one additional measurement since there are now three unknowns: x, y and t_m for the mobile device. Consider the situation in Figure 4.6:

At time t_m the mobile device transmits a signal which is received by the fixed devices A, B and C at t_{ma}, t_{mb}, t_{mc} respectively.

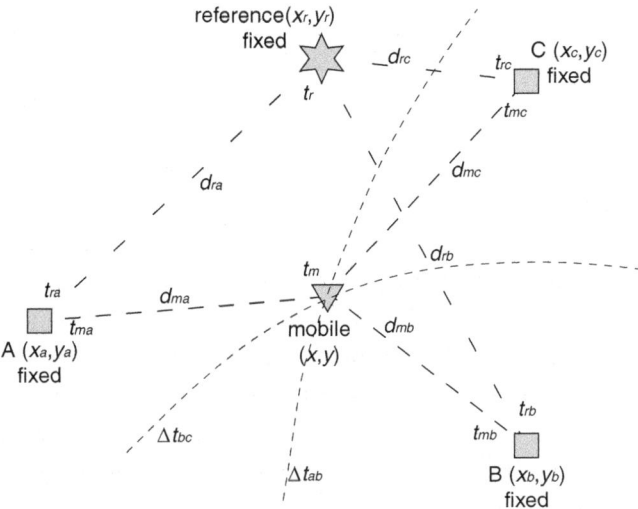

Figure 4.6 Basic TDOA architecture

Consider just A and B for the moment, and assume that their clocks are synchronised, we get the following pair of equations:

$$d_{ma} = c(t_{ma} - t_m) = \sqrt{(x - x_a)^2 + (y - y_a)^2}$$
$$d_{mb} = c(t_{mb} - t_m) = \sqrt{(x - x_b)^2 + (y - y_b)^2} \quad (4.6)$$

Subtracting the second from the first yields:

$$c(t_{ma} - t_{mb}) = \sqrt{(x - x_a)^2 + (y - y_a)^2} - \sqrt{(x - x_b)^2 + (y - y_b)^2}. \quad (4.7)$$

So using two measurements we have eliminated the unknown transmission time t_m and are left with an equation of two unknowns: x, y. There are many positions that can satisfy this equation, the locus of which describes a hyperbola as indicated by the dotted line Δt_{ab} in Figure 4.6. Doing the same for a second pair of fixed devices, B and C yields a similar equation:

$$c(t_{mb} - t_{mc}) = \sqrt{(x - x_b)^2 + (y - y_b)^2} - \sqrt{(x - x_c)^2 + (y - y_c)^2}.$$

$$(4.8)$$

Solving these two equations for the unknown position (x,y) is now a simple matter.

The solution presented here shows three fixed devices measuring the signal transmitted by a mobile device. Clearly the roles can be reversed and the mobile device could be measuring signals transmitted by the fixed devices. In this case the measurement process needs to account for the fact that the fixed devices probably won't transmit their signals at the same time. This is relatively easy to do and many alternative methods of synchronising transmissions or accounting for different transmission times are described in the literature.

4.5.2 Synchronising fixed devices

In the basic description in the preceding section, it was assumed that the fixed devices have synchronised clocks, in order that their time-of-arrival measurements can be directly combined in the maths. There are several ways that this can be done:

1. They may be connected to a wired network, comprising either copper wires, coaxial cables or fibre optic cables. Signals transmitted over this wired network can be used to synchronise the clocks. This may be done using techniques like NTP (Network Time Protocol) or IEEE1588 Precision Time Protocol. With a well-controlled clock source on the wired network it is possible to achieve adequate accuracy of time synchronisation.
2. They may be connected to GPS time transfer receivers. GPS time transfer is able to achieve absolute time synchronisation to around 30 ns or better. Depending on the application requirements this may be good enough.
3. A Local Measurement Unit (LMU) can be used – or in the reverse application described previously these would be reference signal sources rather than LMUs. Since this is a common technique used, and

lower cost than the two preceding options, the principles behind it will be described in more detail.

Figure 4.6 includes a fixed reference source located at a known position (x_r, y_r) and which transmits a signal at time t_r, received by the three fixed receivers at times t_{ra}, t_{rb} and t_{rc}. Since the positions of the reference source and fixed receivers are all known the distances separating them, d_{ra}, d_{rb} and d_{rc} are also known. This means that we can use measurements of the reference source to determine differences between the clocks of the fixed receivers.

So since the clocks at the receivers are not synchronised, let's assume that they have fixed offsets relative to one another. Relating A and C to B:

$$t_a - t_b = \tau_{ba}$$
$$t_c - t_b = \tau_{bc}. \tag{4.9}$$

Therefore the measurements of the time of arrival of the signals from the reference transmitter become:

$$(t_{rb} + \tau_{ba}) - t_{ra} = \frac{d_{rb} - d_{ra}}{c}$$

$$\tau_{ba} = t_{ra} - t_{rb} + \frac{d_{rb} - d_{ra}}{c} \tag{4.10}$$

and

$$\tau_{cb} = t_{rc} - t_{rb} + \frac{d_{rb} - d_{rc}}{c}$$

which we can substitute back into Equations (4.7) and (4.8):

$$c(t_{ma} - t_{mb} - \tau_{ba}) = \sqrt{(x - x_a)^2 + (y - y_a)^2} - \sqrt{(x - x_b)^2 + (y - y_b)^2} \tag{4.11}$$

$$c(t_{mc} - t_{mb} - \tau_{bc}) = \sqrt{(x - x_c)^2 + (y - y_c)^2} - \sqrt{(x - x_b)^2 + (y - y_b)^2} \tag{4.12}$$

and then solve as before to find the unknown position (x,y) of the mobile device.

In practice the clocks in the fixed devices do not have a constant offset, it actually varies with time. Therefore since the measurements from the fixed reference transmitter are received at a different time from those of the mobile device it is necessary to correct for this. Provided the measurements are made within a short space of time relative to the stability of the clock oscillators (typically seconds to minutes) it is usually adequate to represent the clock offset as a simple linear function of time comprising an offset and a drift rate.

$$\tau(t_1) = \tau(t_0) + \delta_\tau(t_1 - t_0). \tag{4.13}$$

Each new measurement of the offset is used to update the model parameters ($\tau(t_0)$ the offset at t_0 and δ_τ the drift rate), typically using a phase and frequency locked loop. This model also acts as a filter to reduce measurement noise.

4.5.3 Alternative approach for solving the equations

The diagram and analysis in the previous section have shown a direct solution for the position using the minimum number of measurements required for the 2D problem. For the 3D case the approach is identical, but one more measurement is required – 4 for the TDOA solution.

In practice one always strives to have more than the minimum number of measurements available and, therefore, instead of using the minimum 3 measurements from fixed receivers it is desirable to use as many as possible. Significant improvements can be obtained with definite advantages to using at least 6 or more with up to about 12 or 16 measurements before the incremental value of adding measurements is outweighed by the complexity of the system and solution.

Having more measurements than required can improve the overall accuracy and it also allows the system to detect, and mitigate, outlying measurements which may be completely erroneous due to measurement error or gross environmental distortions.

One way of solving the overdetermined set of equations is to construct a 'cost' function and numerically minimise the overall solution cost using a suitable minimisation algorithm.

Suppose we have n measurements made by n fixed receivers of the mobile device's signal, we can write the distance between the mobile and the ith receiver as:

$$R_i = \sqrt{(x - x_i)^2 + (y - y_i)^2} \qquad (4.14)$$

and the measured distance from the time observations as:

$$r_i = c(t_i - \tau_i - t_m). \qquad (4.15)$$

These are nominally equal except that the observations are not perfect so in practice there is noise (observation error) to be accounted for. To find the best possible solution we wish to minimise the total measurement error. One way to do this is to find the minimum of the error squared function. Therefore we will use a suitable numerical minimisation technique to find the minimum of the following error (cost) function:

$$F_c(x, y, t_m) = \sum_{i=1}^{n} [R_i - r_i]^2. \qquad (4.16)$$

Provided there are at least 3 measurements from 3 geographically distinct fixed receivers a minimum error (cost) can be found. Provided the equation set is overdetermined (at least 4 or more measurements) the resulting minimum error squared result may be partitioned amongst the individual measurements – these are called the measurement residuals – from which those contributing the largest error can be identified.

As will be appreciated it is possible to develop algorithms and strategies to detect and mitigate rogue or outlying measurements. It is also possible to develop more complex cost functions in which the quality of the measurements used in the equation set is taken into account, thereby assigning more 'value' to clean good measurements and less value to those detected as noisy or potentially erroneous. Since most such algorithms and methods are highly technology and application specific it is left as an exercise for the reader to develop appropriate extensions to his or her requirement.

There are several different numerical minimisation algorithms that can be used, such as, for example, the Broyden–Fletcher–Goldfarb–Shanno (BFGS) method. Press *et al.* [24] includes a number of numerical algorithms that could be used.

4.6 Measuring signal arrival time

4.6.1 Cross-correlation of broadband signal

The maximum likelihood (ML) estimate of the arrival time (delay) is obtained by finding the time offset τ for which the correlation function of the received signal $x(t)$ with the transmitted signal $s(t)$ is maximised.

$$r_{xz}(\tau) = \frac{1}{T_0} \int_{T_0} x(t)s(t-\tau)dt. \qquad (4.17)$$

T_0 the integration time is set as half the duration of the sampling period of the signal.

The signal must be sampled above the Nyquist frequency of the channel bandwidth for the best results.

The correlation may be performed using a sliding correlator or a matched filter.

The sliding correlator is generally used when time offset measurements are relatively infrequent and correlation can only be performed against a small portion of the transmitted signal – such as a training sequence. This is the technique that is often used in mobile cellular systems for example.

A matched filter is generally used when more frequent or regular time offset measurements are needed. This is what is commonly used in UWB (ultrawideband) systems.

It can also be operated as a tracking filter in which the coefficients are updated to match the transmitted signal continuously with the clock to the filter being dynamically adjusted to track the time offset, typically by monitoring early and late filter outputs. This is the method commonly used in GPS receivers.

Given that for many systems the signal sampling rate yields only very coarse time resolution it is usual for the correlator output to be interpolated to determine a more precise time offset.

Consider a relatively narrow band signal of 5 MHz sampled at 10 MHz. This translates into 100 ns sample time resolution which gives a distance granularity of around 30 metres. Therefore the correlation output is normally upsampled or interpolated to 1 GHz or more. 1 GHz gives a distance granularity of about 0.3 m.

The most common interpolation method is a parabolic fit, which is simple but suffers from bias errors. Alternatively a cosine fit is sometimes used, or more advanced reconstructive interpolation methods. Very often the basic methods are combined with compensation techniques, such as bias compensation, or with the use of matched filters. Different methods yield different performance under good and bad signal noise ratios and when signal 'velocity' components are present. Further description of the techniques is beyond the scope of this book and the reader is referred to resources such as Lim *et al.* [25].

4.6.2 Channel models

In an ideal world the transmitted signal would be received delayed in time and attenuated (smaller amplitude) but otherwise unchanged. However, in practice this is never the case: real signals are subjected to multipath and other effects and sometimes the direct path is not even present in the received signal. For the purpose of positioning it is desirable to measure the direct path, which is usually the earliest arriving signal. Therefore one attempts to determine the first (earliest) correlation peak, and not neces-sarily the largest.

Real-world environments are forever changing and it is very difficult to replicate a specific set of conditions precisely. Therefore development is dependent on using captured data sets and on using channel models to define what happens to the radio signals between transmitter and receiver.

amplitude

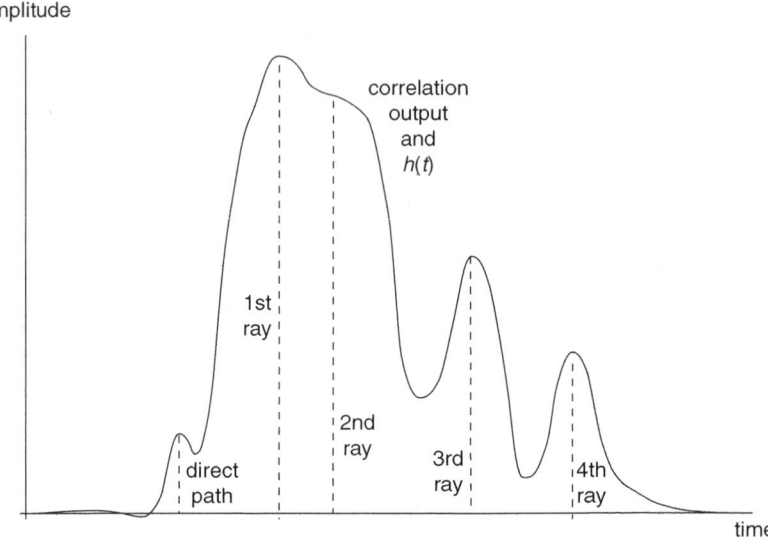

Figure 4.7 Illustration of $h(t)$ and correlator output

A generic channel model is defined thus:

$$h(t) = \sum_{k=0}^{M} h_k \delta(t - T_k). \qquad (4.18)$$

The received signal $h(t)$ comprises a direct path signal and M further replicas each delayed in time by T_k. Each path is subjected to the complex attenuation h_k having both amplitude and phase response.

An illustration of a simple channel response is shown in Figure 4.7.

In this case the direct path is distinct, but much smaller than the subsequent paths. The first and second multipath signals (1st ray and 2nd ray) are masking one another because they are closer together than the bandwidth of the signal allows to be resolved. The challenge for real positioning systems is to be able to reliably and accurately resolve the earliest time of arrival measurement under real and varied channel conditions.

4.6.3 Other advanced techniques

A number of other techniques and algorithms have been developed to try and extract better estimates of the time of arrival of the signal.

Collectively these techniques are sometimes referred to as super resolution algorithms. Most of these techniques attempt to estimate or construct the channel impulse response from its frequency response.

The basic relationship is obtained by representing the impulse response in the frequency domain:

$$H(f) = \sum_{i=0}^{M} a_i e^{-j2\pi\tau_i f}. \qquad (4.19)$$

Therefore the simplest way of obtaining the channel impulse response from its frequency response is to perform an IFFT (inverse fast-Fourier transform). As per usual practice this is normally done after windowing the input data so as to reduce side lobes in the output. The earliest measurable time of arrival can then be obtained by looking for the leading edge of the first peak in the impulse response.

Another popular super resolution algorithm is MUSIC (Multiple Signal Classification) which splits $H(f)$ into signal and noise components in an attempt to estimate the individual multipath elements. It starts with an estimated number of multipath rays and uses the eigenvectors of the autocorrelation matrix of $H(f)$.

Another class of algorithms includes ESPRIT (Estimation of Signal Parameters via Rotational Invariance Techniques) and 'Matrix Pencil'. They attempt to reconstruct $H(f)$ using a finite number of complex sinusoids.

Humphrey [26] provides a concise overview of the major super resolution algorithms used to measure time of arrival.

A number of algorithms for joint estimation of both time and direction of arrival are also being developed. In addition to ESPRIT, these include: joint angle and delay estimation (JADE) and space-alternating generalised expectation maximisation (SAGE). Li [22] provides a comparative assessment of the performance achieved by each.

4.6.4 Phase measurements

Phase can also be used to measure time of arrival. Very simply the signal's phasor rotates at the signal frequency, which means that due to the

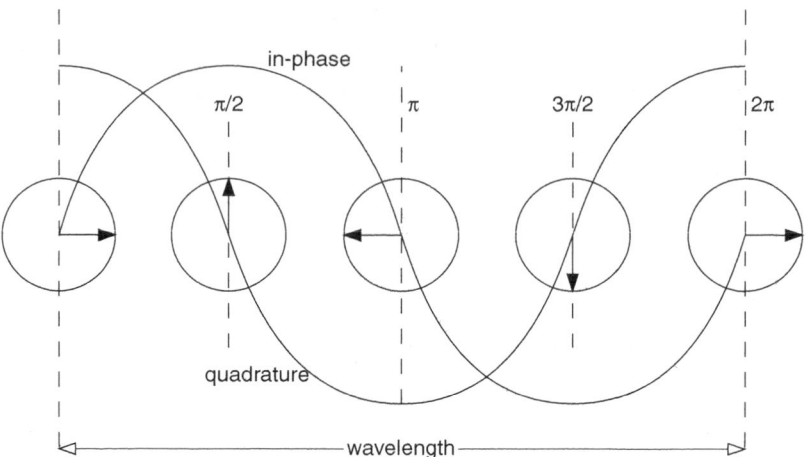

Figure 4.8 Distance as a measure of phase

wavelength and speed of propagation relationship it rotates by one cycle, 2π radians, every wavelength. Therefore by measuring the phase difference from transmitter to receiver one has an estimate of the distance the signal has travelled modulo wavelengths (Figure 4.8).

Phase measurements can be easily made to high accuracy – it is relatively straightforward to measure to an accuracy of 1% of the wavelength. However, phase measuring systems present a number of challenges of their own:

- The wavelength cyclic ambiguity means that an unknown number of cycles needs to be resolved for a full solution. At 1.5 GHz the wavelength is 0.2 m (20 cm), which means that we can resolve millimetres of distance, but we need to know the distance to better than 20 cm before this measurement is useful. Lower frequencies offer a better trade off – for example 3 MHz has a wavelength of 100 m and should yield a precision of better than 1 m. Unfortunately transmitting radio signals at 3 MHz is not practical or legal in most cases.
- The initial phase condition of the transmitter is hard to determine and therefore one has to deal with an unknown initial phase condition. There are ways of resolving this unknown offset, for example by using a local measurement device at a known location.

- Resolving multipath in phase measuring systems is notoriously difficult. Two different copies of the signal arriving at the receiver with different time delays sum to give a signal with phase offset being a weighted sum of the two, and it is not possible to distinguish between them without resorting to secondary techniques.
- Phase measuring systems are narrow band and as such they can suffer from frequency selective fades, which in the worst case could cause complete cancellation of the received signal.

Despite these challenges, phase measuring systems are very useful for obtaining the highest possible precision in many positioning systems. The use of carrier phase in GPS is the fundamental principle on which ultra high precision survey and RTK systems operate.

They are also useful for many general purpose radiolocation systems. Suppose for example that instead of measuring the signal carrier, one measures two carriers (say two carriers in an OFDM multiplex) spaced apart by a frequency Δf. Since these are generated concurrently and therefore have a known time when both have a known initial phase condition, the problem of the unknown initial phase is solved. Furthermore, by using the difference between the two we effectively have a phase measurement related to the wavelength of the difference frequency. So suppose the two carriers were spaced apart by 3 MHz, this leads to a wavelength of 100 m and 1 m or better measurement resolution – independent of the actual radio band in which the signal is being transmitted.

4.7 Positioning using cellular mobile networks

The most important cellular mobile standards today are: GSM, WCDMA, LTE, CDMA2000 and WiMAX. The first three are standards developed by 3GPP (3rd Generation Partnership Programme), CDMA2000 is developed by 3GPP2 (3rd Generation Partnership Programme 2) and WiMAX is the interoperable implementation of the IEEE 802.16 standards ratified by the WiMAX Forum. All of these standards include many different techniques and protocols for positioning and device localisation, so instead of presenting the methods used on a standard-by-standard basis this section describes the major techniques used and their applicability in the different standards.

4.7.1 Background to mobile cellular positioning

Whilst it has always been envisaged that a rich set of commercial Location-Based Service (LBS) applications would be developed for mobile devices, the original impetus for developing positioning capability in the networks was for the localisation of emergency calls. In the USA this was formulated as a legal requirement on all cellular network operators to be able to pinpoint the location of a device making an emergency call – known as the E911 (Enhanced 911) mandate. E911 stipulates an accuracy, probability and maximum time to obtain this information.

For network-based technologies E911 must be accurate to 100 m 67% and 300 m 95%, and for terminal-based methods it must be accurate to 50 m 67% and 150 m 95%, both delivered to the responding service within 6 minutes of the call origination.

In Europe a similar directive called E112 (the emergency number is 112) has been introduced, although E112 does not stipulate a particular accuracy; it merely requires 'best effort'. In some respects this is a pragmatic solution to the non-trivial task of specifying and validating accuracy figures for devices in general use by consumers in the real world.

Several other countries have introduced or are planning to introduce similar emergency call localisation regulations.

However, the real future benefits may well lie in commercial applications that can make use of accurate and trustworthy device positioning information.

4.7.2 Satellite positioning

4.7.2.1 Assisted GPS

All of the mobile cellular networks specify the use of GNSS, very often as the primary positioning technology. In almost all cases an Assistance service is defined and specified.

For most devices the GPS receiver is turned off most of the time in order to conserve battery power. Therefore, when a position is requested the GPS receiver has to go through the startup procedure of acquiring the satellites and receiving and decoding the almanac and ephemeris data from the signals. From a cold start this process can take several minutes, assuming that the satellite signal is available in the first place. This is called the Time To First Fix (TTFF).

Assisted GPS is an architecture in which a GPS reference receiver accessible by devices in the cellular network is operated by the service provider. This reference receiver continually monitors all the satellites in view and captures the satellite information. The mobile device can request, via a network data service, the satellite data from the reference receiver, thereby bypassing the need to acquire the satellite signal and decode the data transmitted by it. Having the satellite data to hand then allows the mobile device to predict far more accurately which satellites to search for and leads to much more rapid acquisition of the signals. An assisted GPS receiver may be able to acquire the satellite signals and compute a position within about 15 seconds or less from cold start – assuming that the satellite signals are receivable by it.

4.7.2.2 Other GPS aiding techniques

In addition to the exchange of almanac and ephemeris data for the purpose of aiding signal acquisition there are other methods that have been developed to assist or support GPS positioning.

Knowing exactly which satellites to search for goes some way towards acquiring the signals, but the receiver still does not know the precise time or frequency offsets.

Since the mobile terminal is operating within a cellular network which is operated to high precision it is able to track the network clock which provides very good frequency accuracy. Therefore the relatively poor frequency accuracy of its own local clock can be overcome. Given a rough position for the mobile device (nearest Cell position) and knowledge of satellite positions, the Dopplers can be estimated and as a result the frequency search space for satellite acquisition can be significantly reduced.

Narrowing the time search window is slightly trickier. In those networks operating synchronised base stations (Node Bs), such as CDMA2000 and WiMAX, the base stations are normally synchronised using GPS time transfer. The mobile device therefore has a sense of absolute time, delayed by the unknown propagation distance from its serving cell. In most cases synchronisation to a number of microseconds or tens of microseconds can be achieved. In those networks operating unsynchronised base stations (GSM, CDMA) acquiring time synchronisation to the level of a few microseconds is not so easy. Rowe [27] describes a method of using network signals for the purpose of maintaining accurate clock synchronisation to aid GPS signal acquisition.

Having very good time and frequency alignment narrows the satellite search and makes it significantly easier and quicker to acquire satellite signals under less favourable conditions.

4.7.3 Cell and proximity methods

The most basic solution is to simply report the identity of serving cell or base station for the mobile. Its location is then taken as the position of the cell as obtained from the network. Depending on the size of the cell this leads to position estimates which can be anything in the range of 100 m to 10 km or more in error. Usually each base station is configured with a number of sectors, three being common. In this case the position can be assigned to the centre of the areas covered by the sector in which it is found.

There are a number of ways to improve the most basic form of cell ID:

1. Using signal strength measurement (RSSI). This can be used to obtain a coarse estimate of the distance from the serving cell. In CDMA networks with power control it may also be necessary to take into account the current transmit power.
2. A coarse measurement of round-trip time may be available. In GSM networks this could be obtained from the timing advance correction, and in most other networks similar timing compensation parameters

are used to improve the time alignment between the base station and mobile device.

3. Measurements of several base stations are likely to be possible. In the simplest situation this is merely a list of neighbouring cells and their signal strengths.

Cell ID is usually defined as a fall back method in the mobile cellular standards and would be used when GPS and other higher precision methods are not available. Cell ID is the lowest common denominator and is always available as long as the mobile device is connected to the network and has service.

4.7.4 Terminal-based time-of-arrival techniques

Time-of-arrival techniques, usually time difference of arrival, are often considered as the first fall back method in the event that GPS is unavailable, although in many cases OTDOA (observed time difference of arrival) provides a better solution than GPS.

Generally these techniques fall into two classes: UE (user equipment) based and Network based. The advantage of network-based solutions is that the mobile device does not need to do anything and therefore standard unmodified mobile terminals can be located. However, these systems usually incur additional cost and complexity in the network. There are also various in-between configurations in which the terminal or network obtains assistance from the other and is not, therefore, fully independent. Network-based methods are described in the next section.

4.7.4.1 E-OTD, OTDOA and similar methods

E-OTD (Enhanced Observed Time Difference) is the technique specified for GSM and OTDOA (Observed Time Difference of Arrival) is a more general name for the use of observed time difference in other standards including WCDMA and LTE.

The mobile terminal device observes and measures signals from multiple base stations (Node Bs). In GSM this is done using a training sequence' at the beginning of the broadcast frame and in WCDMA the CPICH (common pilot channel) is measured. Time offsets (differences) between the signals from the different base stations are measured and reported to a server in the network which is able to use them to compute the position of the mobile terminal given knowledge of the locations of each base station, and the time offsets between their clocks. The time offsets between base station clocks are obtained using LMUs (Local Measurement Units) positioned at known locations throughout the network. They are often co-located with the base stations. The process for calculating the mobile terminal position is as described earlier in Section 4.5.

In the case of WCDMA networks base stations share a common radio channel and the signals are extracted by the terminal based on code correlation. For the serving base station, which is usually the strongest signal, this is easily done, but for more distant base stations the weaker signals can easily be masked by nearer stronger ones. This means that the ability to receive signals is often limited by the presence of stronger closer signals rather than the performance of the radio. This has the effect of 'deafening' the receiver to the weaker signals and limits the number of neighbours that can be measured. A number of techniques for dealing with this problem and improving the 'hearability' of distant base stations have been developed.

Within 3GPP a technique called IP-DL (Idle Period on the DownLink) has been standardised. This technique provides for base stations to insert short periods of 'silence' in their downlink transmission, by turning off the transmitter for a short period of time. This short interruption has little effect on the link performance, but during this period the mobile terminal is able to measure weaker more distant signals, and thereby overcome some of the signal masking. The standards provide for different ways of managing the insertion of idle periods: randomly or synchronously.

Duffett-Smith [28] describes another technique called 'cumulative virtual blanking (CVB)', that offers superior performance to IP-DL, but with some added complexity.

4.7.4.2 OTDOA without LMUs
The standardised architecture for OTDOA (and E-OTD) makes use of LMUs (Local Measurement Units) which enable the clock offsets

between base stations to be measured. Drane [29] describes a method of using measurements sourced from a number of mobile devices at unknown locations, to solve for the time offsets in the network and therefore enable operation without the need for LMUs. Apart from their use in TDOA solutions, knowing the time offsets is also a valuable ingredient of high performance assisted GPS systems.

4.7.5 Network-based time difference of arrival techniques

The advantage of network-based techniques is that the mobile terminal plays a passive role and does not need to actively participate in the positioning process. As such the position of any mobile terminal can be measured by the network.

4.7.5.1 U-TDOA: uplink TDOA

Uplink TDOA is another technique standardised by 3GPP for computing mobile terminal positions.

The network is populated with a number of LMUs. When the mobile device transmits a signal to the network the time of arrival of the signal is measured at the serving base station and at least two LMUs. The time differences are computed and the position of the mobile device is solved as described for general TDOA.

In order to measure the signal from the mobile device it needs to be active. The measurements can either be made when it is in a call (such as a 911 call), or it can be stimulated to transmit.

4.7.5.2 PCM: pilot correlation method

PCM is a fingerprint technique for UMTS based on CPICH (Common PIlot CHannel) power at the mobile device. CPICH power is measured for all Node Bs within range and characterised throughout the area with the values being modelled and stored in a database.

The received power is measured and used by the mobile device for handover and network management functions, so all that is required is for the measurements to be reported to the network. They are matched against the database and the most likely position estimated.

A similar fingerprint method for GSM, called DCM (database correlation method), uses signal strength measurements of the serving cell and six neighbouring cells. These measurements are collected and reported by the GSM handset as part of its network management report.

As discussed previously fingerprinting methods can lead to fairly good results provided that nothing changes in the environment. Unfortunately the environment is ever changing, plus mobile handsets are carried and held by users in many different orientations and places, so, for example, the measurement obtained by a phone clipped to one's belt is very different from one in a pocket or in a handbag. With this variability, these methods should be considered at best approximate.

4.7.6 SUPL: secure user plane localisation

Almost all of the mobile cellular standards have specified signalling protocols for Assisted GPS, Cell-ID, various TDOA techniques and other localisation techniques. These require the network infrastructure and user terminals (mobile devices) to support the signalling protocols and for the network operator to provide an assistance server. An alternative approach standardised by the OMA (Open Mobile Alliance), called SUPL (Secure User Plane Localisation) has been developed for exchanging positioning related data via a general purpose data connection using IP (Internet Protocol).

The advantage of SUPL's user plane approach is that any network capable of supporting a data connection between the user and the outside world (Internet) could be used for SUPL and it is not necessary for the network and user device to support the standardised control plane protocols.

When a mobile terminal wishes to determine its position and requires additional information from the network it opens an IP connection to an appropriate SUPL server, requesting the additional network assistance information. It computes its position and may, if required, send the result to a SUPL compliant location application.

Conversely if a network application requires the position of a terminal it may send an initiating request to the terminal, using a suitable push service (such as SMS). The terminal opens the requested data connection which is used to exchange positioning related information until the position has been calculated and delivered.

SUPL also includes standards for exchanging position related data between different SUPL servers and between the servers and mobile networks.

4.8 Ultrawideband (UWB)

4.8.1 Introduction to UWB

Ultrawideband signals are characterised by their very large bandwidths compared to conventional signals such as Wi-Fi, cellular mobile or TV and radio broadcast signals. The FCC defines UWB signals as having > 500 MHz bandwidth or a fractional bandwidth > 20%, where:

$$B_{frac} = \frac{f_h - f_l}{f_c} \tag{4.20}$$

where f_h and f_l are the -10 dB upper and lower frequency points and f_c is the carrier frequency (Figure 4.9).

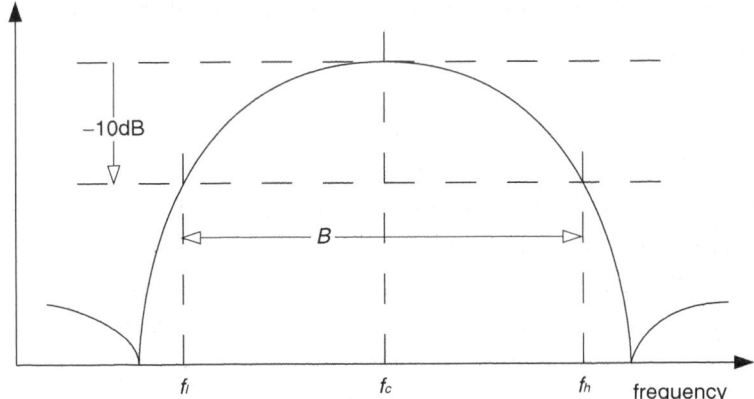

Figure 4.9 UWB spectral power illustration

Since UWB systems occupy such a large bandwidth they typically coexist with other radio systems by transmitting at power levels much lower than the other systems. The regulatory framework for UWB is still evolving.

4.8.2 Regulatory framework for UWB

The FCC has published regulations governing the use of UWB. In general UWB systems must transmit at power levels low enough not to interfere with other systems occupying the band. Specifically they are permitted to transmit such that the power spectral density between 3.1 GHz and 10.6 GHz does not exceed −41.3 dBm measured in a 1 MHz band. Outside of this band power levels vary and generally need to be lower. In particular the GNSS bands from 960 MHz to 1.61 GHz may not be more than −75 dBm.

The EU regulations look rather different. Specifically:

- In the band 6.0 GHz to 8.5 GHz allowed power is −41.3 dBm/MHz (same as FCC)
- 3.1 GHz to 4.8 GHz: low duty cycle operation is required but −41.3 dBm/MHz is permitted
- generally −70 dBm and lower outside of these bands
- −85 dBm below 2.7 GHz and −90 dBm below 1.6 GHz.

In Japan the core band is 7.25 GHz to 10.25 GHz at −41.3 dBm/MHz; 4.8 GHz to 6.0 GHz is excluded with out-of-band levels being similar to the EU regulations.

4.8.3 UWB signal structures

The most common form of UWB is impulse radio (IR UWB) in which data is transmitted as very short pulses in which the pulse width is a fraction of a nanosecond and the interval between pulses is long compared to the pulse width. Encoding is either through the pulse polarity, or pulse position (time relative to standard pulse interval).

However, within the IEEE 801.15.3a standard two alternative physical layer specifications have been defined:

- DS-UWB, using direct sequence spread spectrum, defined for operation in bands 3.1 GHz to 4.8 GHz and 6.2 GHz to 9.7 GHz.
- MB-OFDM UWB, a multiband OFDM implementation defined around 14 sub-bands each 528 MHz wide. The sub-bands are divided into band groups with, typically, three sub-bands in each which are used on rotating basis. An MB-OFDM device must support at least one band group.

Although not strictly UWB according to the FCC definition, it is worth mentioning the IEEE 802.15.4a physical layer specification for 2.4 GHz ISM band operation based on CSS (chirp spread spectrum) which is defined to use the entire 80 MHz bandwidth available. It uses swept chirps in which the signal is swept either up or down in 1 μs, the polarity of the sweep being used to indicate data encoding.

4.8.4 UWB for localisation

Given the very wide bandwidths in UWB systems (as much as 6 GHz, but a minimum of 1.5 GHz) they are ideally suited for determining position. At 1.5 GHz we have the equivalent of about 0.67 ns in time, or about 0.2 m in distance. Under ideal conditions UWB systems can resolve 1 cm or better in range, and the large bandwidth allows multipath rays differing by as little as 20 cm to be distinguished from one another.

Apart from the physical signals and the systems for generating, receiving and measuring them, UWB positioning systems use the same fundamental techniques discussed earlier (range measuring, time-of-arrival and time-difference-of-arrival) to compute positions.

For further information and detail about UWB systems for positioning Sahinoglu [30] is a great starting point.

4.9 Other radiolocation systems

There are many other radiolocation systems – historical, current and emerging. Historical systems include DECCA, LORAN and others. Systems using other existing radio networks, such as radio or TV transmitters have been built, and there are many other commercial systems in existence. Samama [1] provides a good historical overview.

5   Inertial navigation

5.1 Principle of inertial navigation

Inertial navigation is based on Newton's first law of motion: '*Every body remains in a state of rest or uniform motion (constant velocity) unless it is acted upon by an external unbalanced force*'. This means that in the absence of a non-zero net force, the centre of mass of a body either remains at rest, or moves at a constant speed in a straight line. Inertia is the property of a body that links force with mass and change of velocity; it applies to both linear motion and rotation.

Therefore if we can measure the forces acting on a body it is possible to convert these into a change of motion state using the laws of motion, provided we know the initial conditions – position and velocities. By continuously integrating the changes in motion state it is possible to track the resulting position and velocity.

If instead of measuring the force of the object we measure its acceleration – given that:

$$F = ma \tag{5.1}$$

this means that we don't need to know the mass of the object, and given an acceleration we can compute the velocity by integrating once and the distance travelled by integrating a second time, assuming that we know the initial conditions.

Consider a very simple one-dimensional example in which an object starts from rest and then travels in a straight line. If we measure the acceleration a we can compute the velocity and distance travelled over time given initial conditions v_0 and d_0.

$$\begin{aligned} v &= \int a\,dt + v_0 \\ d &= \int v\,dt + d_0 \end{aligned} \tag{5.2}$$

Extending this to the general 3-dimensional case in which initial conditions may be uncertain and the sensors are not ideal is far from trivial, but with the development of MEMs (micro-machined electromechanical systems) sensors and access to abundant computing power the use of inertial data is within reach of many positioning systems.

The big advantage of inertial navigation is that it does not rely on any surrounding infrastructure or external radio signals, and as such falls into the general category of 'dead reckoning' systems. This is also one of its major weaknesses: over time errors accumulate and the position or velocity of the object in relation to its environment becomes less certain.

Before getting into the detail of how such systems work, we should just remind ourselves about coordinate systems and in particular what an inertial coordinate frame is.

An inertial coordinate frame is one in which Newton's laws of motion hold true: they are neither rotating nor accelerating. In Chapter 2 we learnt about coordinate frames: geographic coordinate frames of longitude and latitude attached to the Earth; and Cartesian coordinate frames that could be attached to the Earth (ECEF) or a local area or to a moving object or which could be completely independent. Since the Earth is moving through space, circling around the Sun and rotating on its axis, any coordinate frame attached to the Earth is, by definition, not an inertial coordinate frame. Therefore in order to solve the inertial navigation problem we will need to adopt a suitable inertial coordinate reference frame. Depending on the nature of the problem being solved and the sensitivity of the sensors being used it is often adequate to adopt a reference frame centred on the Earth but orientated in a fixed arrangement relative to the stars – and therefore not rotating with the spin of the Earth.

The most common approach to inertial navigation is called 'strapdown' inertial navigation. In this arrangement a set of three orthogonal accelerometers and three orthogonal rate gyroscopes are attached to the body frame of the object being tracked. The accelerometers are used to measure translational accelerations – the result of unbalanced forces acting on the body – and the rate gyroscopes are used to measure the rate of rotation of the object. Given a known initial condition: position, velocity and orientation, the sensor measurements are used to track the object's motion and compute subsequent positions, velocities and orientation. Since the sensors are coupled to the body axes, their measurements represent the movement of the body axes of the object in the inertial coordinate frame, which can be converted back into a useful real-world coordinate frame given knowledge of how that frame moves relative to the inertial reference frame.

Strap-down inertial navigation is not the only possible way of arranging the sensors for inertial navigation, the earliest systems used platform sensors in which the sensors were completely isolated from the vehicle using a set of gimbals, but in this book we will concentrate on strap-down techniques.

5.2 Sensors used for inertial navigation

5.2.1 Accelerometers

The big advantage in measuring acceleration is that it can be measured without reference to any external infrastructure or devices. In general terms an accelerometer is a device in which a proof mass is mounted in such a way that the force acting on the mass is measured by the sensor. This allows the sensor to determine the acceleration of the object to which it is attached without knowing the mass of the object or the force acting on it as a whole.

Since most of the navigation we're concerned about in this book is done in the context of real-world applications we also need to take into account the force of gravity. Since gravity also acts on the proof mass in the accelerometer, the acceleration measured by the accelerometer includes a component of gravity. An accelerometer measuring vertical acceleration will measure the full gravity value, which at sea level is approximately $9.81\ ms^{-2}$, and therefore the acceleration reported by it will be:

$$a = f + g \tag{5.3}$$

where g is the gravitational value and f is the non-gravitational force acting on the accelerometer. Being able to offset the gravity vector is an essential part of the inertial navigation solution, and ways of doing this will be described later in Sections 5.3 and 5.4.

Early accelerometers were mechanical devices, but recently solid state and MEMS accelerometers have been gaining popularity as their performance improves. For a detailed discussion of different types of accelerometers see Titterton and Weston [31]. For mass market applications the use of MEMS devices is particularly attractive because they consume very little power, are small and inexpensive. However, most consumer grade MEMS accelerometers do not give good enough performance for use in a true inertial navigation context.

In practical strap-down navigation systems a set of three accelerometers is usually mounted on the object with their respective measurement axes orthogonal to one another and aligned in a known arrangement with respect to the object's body axes.

Accelerometers are subject to errors and it is worth noting some of the important error sources (see Figures 5.1 and 5.2 for illustrations):

- Bias or zero offset error – displacement from zero in the absence of any applied acceleration;
- Scale error – error in the ratio of output value to applied acceleration;
- Cross-coupling errors – measurements induced by accelerations in axes orthogonal to the accelerometer's measurement axis;
- Temperature effects, that can cause bias, scale and other errors to change with changes in temperature;
- Non-linearity – output variation in which the output does not vary exactly linearly with the applied acceleration;
- Alignment errors in which the measurement axis of the accelerometer is not perfectly aligned with the desired axis;
- Offsets between the three measurement axes of the accelerometers and the body axis coordinate frame of the object.

5.2.2 Rate gyroscopes

Gyroscopes are used to either measure an angle (displacement gyroscope) or the angular rate of rotation about an axis (rate gyroscope). Generally

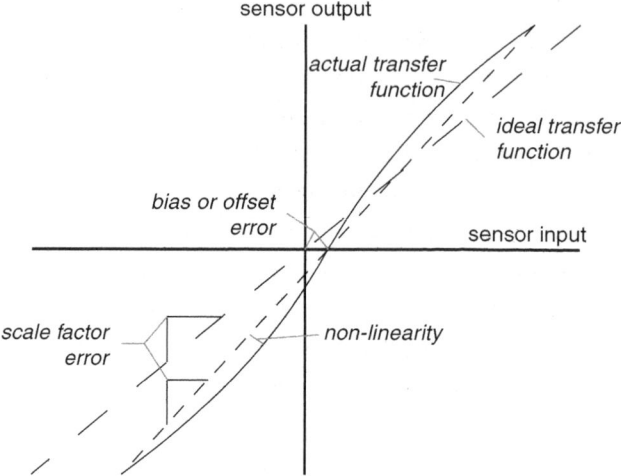

Figure 5.1 Illustration of sensor transfer function errors

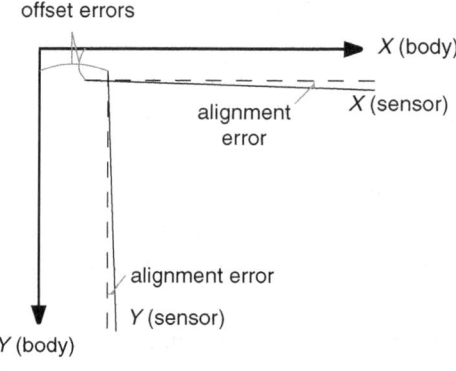

Figure 5.2 Illustration of sensor misalignment errors

displacement gyroscopes are used in stabiliser platforms, whereas for strap-down inertial systems rate gyroscopes are usually used, and we will simply refer to them as gyroscopes.

There are many different types and constructions of gyroscopes and the reader is referred to Titterton and Weston [31] for a good description of the different kinds of gyroscope sensors and how they work.

In reality many modern sensors, such as MEMS gyroscopes, are not really gyroscopes since they do not rely on the dynamic behaviour of a rotating body; however, we will refer to all sensors that measure angular rates as gyroscopes for the purposes of this discussion.

In strap-down inertial navigation the gyroscope is used to measure rotation about an axis. Typically three gyroscopes are arranged orthogonally in a fixed relationship with the body axes of the object being tracked. The gyroscopes are used to keep track of the orientation (attitude) of the object (in reality, the body axes of the object), by integrating the rate of rotation to give an angle, so that combined with measurements from the accelerometers the position and orientation can be computed over time given a known starting point.

Like accelerometers gyroscopes are not perfect devices and suffer from a number of errors that need to be taken into account in the navigation solution:

- Bias or offset, also called drift, is the angular rotation rate reported in the absence of any rotation input;
- Scale factor error arises when the output to input ratio varies from nominal;
- Cross-coupling errors arise when rotation about orthogonal axes results in a change to the measured rate output;
- Acceleration dependent errors can arise if the output varies with linear acceleration of the device;
- Temperature effects can cause changes to bias, scale and other errors as temperature varies;
- Non-linearity in which the change in sensor output does not vary linearly with change in applied rotation rate;
- Alignment error in which the measurement axis is not perfectly aligned with the desired measurement axis;
- Offsets between the measurement axes and the body axis coordinate frame result in measured rotation rate about a point not coincident with the origin of the body axes.

Since the gyroscope measures the rate of rotation about an axis, we have adopted the usual convention of the right-hand rule describing positive rotation about the axis. This is illustrated in Figure 5.3.

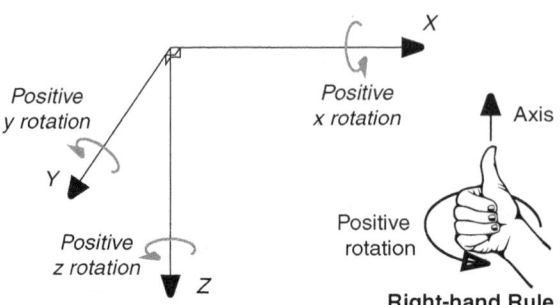

Figure 5.3 Convention for measuring rotation about an axis

5.3 Architecture of a strap-down inertial navigation system

Let's start by assuming that we are going to navigate relative to an inertially fixed coordinate frame – one that is not accelerating or rotating. Six sensors, three accelerometers arranged orthogonally and three rate gyroscopes arranged orthogonally, are used to measure motion relative to the inertial reference frame. Whilst there are arrangements for specific navigation requirements in which fewer or more sensors may be used, or in which they are not arranged orthogonally, these specialist applications fall outside the scope of this book and the interested reader is referred to the specialist literature on the subject.

The sets of orthogonal sensors are attached to the object being positioned in a fixed and known alignment with the body axes of the object. This is the reason for describing such a system as a 'strap-down' inertial navigation system – being attached to the object the sensors are actually measuring the motion of the object's body axes relative to an inertial frame and it is knowledge of the position and orientation of the object's body axes that describes its position and orientation.

The rate gyroscopes are used to calculate the orientation of the object. Given a known initial condition and rate of rotation as a function of time, integrating gives the attitude or orientation. Note that this is the orientation of the body axes of the object with respect to our inertial navigation reference frame.

Knowing the orientation of the object allows the accelerations measured by the accelerometers, along the body axes, to be transformed into the inertial reference frame. This is done by multiplying by the rotation matrix obtained using the rate gyroscope measurements.

The specific accelerations in the inertial reference frame are integrated to give velocity, which when integrated a second time gives distance. With known initial conditions: orientation, velocity and position the result of transforming and integrating the accelerometer measurements yields velocity and position estimates in the inertial coordinate frame being used.

Unfortunately it is rare that the practical coordinate frame in use by everyday applications is an inertial coordinate frame. Generally it is attached to the Earth and very often it represents a small part of the surface of the Earth; perhaps a standard map frame based on UTM coordinates or latitude, longitude and height, or sometimes an application-specific local frame aligned with the field of interest. These practical everyday coordinate frames are not inertial since the Earth is rotating on its axis and moving around the Sun. Therefore, the inertial outputs also need to be transformed into the working coordinate frame and since this frame is affected by gravity and other effects such as Coriolis forces as shown in Figure 5.4 it is also necessary to include corrections for them.

The first and most significant correction needed is for gravity. Since almost all of the applications of concern in the context of this book are Earth based the gravitational field needs to be taken into account. The nature of the accelerometer means that it measures the force of gravity which is a vector directed generally towards the centre of the Earth. However, the Earth is not a sphere: it is approximately ellipsoidal, but with an uneven surface and uneven density throughout the body. The ellipsoidal shape means that the vector linking a point on the surface of the Earth and its centre is not likely to be normal to the surface of the Earth. Furthermore the rotation of the Earth causes a further force called the Coriolis force on the accelerometer. The accelerometer measurement reflects the total force it experiences. Given knowledge of the Earth's rotation and one's position, it is possible to correct for both gravity and Coriolis forces in the navigation equations.

As can be appreciated continual integration of the sensor measurements can and does lead to exponentially growing errors in the final position. The smallest error in the initial conditions will eventually grow to be

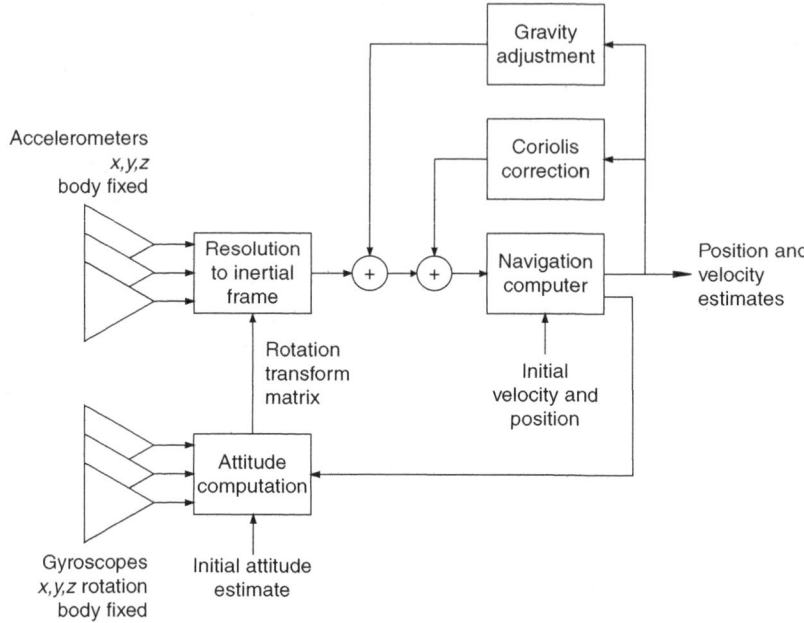

Figure 5.4 General architecture of strap-down inertial navigation system

unacceptably large and the sensor errors will be integrated over time leading eventually to very large output errors. Therefore, starting from an initial known condition the operational period over which the strap-down inertial system is useful is dependent on the accuracy required and the performance of the sensors. In practical systems a means to continually or periodically recalibrate the initial conditions and sensor errors is usually provided, and means for doing this are described later in Section 5.6.

Depending on the accuracy required by the application, the sensitivity of the sensors used, and length of time required for independent operation it may be possible to simplify the equations and in some cases it may even be possible to ignore secondary effects such as the Coriolis force.

5.4 Navigation equations

In this section we summarise a basic set of typical navigation equations. The detailed construction of the equations varies depending on choice of navigation coordinate frame and the extent to which sensor errors and Earth geoid anomalies need to be taken into account.

One of the most common local coordinate frames in use is the local geographic frame in which position is described in terms of latitude, longitude and height, and velocity as north and east velocities. Starting with the fundamental navigation equation:

$$\dot{v}^i = f^i + g^i = C_b^i f^b + g^i \tag{5.4}$$

the specific forces acting on the object result in an acceleration. This equation holds true in an inertial reference frame (indicated by the superscript i). However, the accelerometer measures the specific forces in the body axes (indicated by the superscript b) and these are transformed into the inertial frame by multiplying by the direction cosine matrix C.

However the Earth is rotating relative to the inertial coordinate frame, so skipping over the detailed derivation it can be shown that we obtain the following navigation equation:

$$\dot{v}_e^i = C_b^i f^b - w_{ie}^i \times v_e^i + g_l^i. \tag{5.5}$$

Next we need to express this in terms of our navigation coordinate frame. Once again skipping the derivation it can be shown that the following navigation equation is obtained:

$$\dot{v}_e^n = C_b^n f^b - \left[2w_{ie}^n + w_{en}^n \right] \times v_e^n + g_l^n. \tag{5.6}$$

Here we have the rotation matrix transforming body axis measurements into the navigation frame and two terms representing the sum of the rate at which the Earth is rotating relative to the inertial frame and the rate at which the navigation frame is rotating relative to the Earth fixed reference frame. The gravity vector is represented as the local gravity vector in the navigation frame. Note that the gravity vector comprises the sum of the mass attraction force and the centripetal acceleration caused by the rotation of the Earth. Furthermore the reader will appreciate that some assumptions, or knowledge of, the shape of the Earth are required in order to accurately transform between the local geographic reference frame and the Earth fixed reference frame. These relationships were described in Chapter 2.

A more detailed description of inertial navigation systems is beyond the scope of this book and the reader is referred to the specialist texts on the subject, such as Titterton and Weston [31].

5.5 Brief review of the errors in the strap-down inertial system

There are many sources of errors, particularly when using low-cost MEMs sensors, and most of them drift or vary with time, making one-shot calibration of limited use. To illustrate the magnitude and significance of these errors a typical 0.01 °/s gyroscope bias (fairly typical of current MEMs sensors) alone leads to a position error of more than 60 metres during a one minute unaided navigation interval.

Summarising the relationship between error sources and the position sensitivity as a function of time we have the following major error sources:

$$
\begin{aligned}
\delta p(t) \propto \delta p(t_0) + \delta v(t_0)\Delta t + \delta q_A(t_0)V\Delta t + \delta q_{\sigma,\gamma}(t_0)g\frac{\Delta t^2}{2} \\
+\delta b_a(t_0)\frac{\Delta t^2}{2} +\delta SF_a(t_0)\frac{\Delta t^2}{2} + \delta b_g(t_0)g\frac{\Delta t^3}{6} + \delta SF_g(t_0)\frac{\Delta t^3}{6}
\end{aligned}
\tag{5.7}
$$

$\delta p(t)$	position error after time t
Δt	time since start of unaided navigation
$\delta p(t_0)$	initial position error
$\delta v(t_0)$	initial velocity error
V	average velocity during unaided navigation
g	local gravity constant
$\delta q_A(t_0)$	initial azimuth error
$\delta q_{\sigma,\gamma}$	initial pitch and roll errors
$\delta b_a(t_0)$	accelerometer bias
$\delta b_g(t_0)$	gyroscope bias error
$\delta SF_a(t_0)$	accelerometer scale error
$\delta SF_g(t_0)$	gyroscope scale error

5.6 Integration with other positioning technologies

Since inertial navigation systems built using accelerometers and rate gyroscopes drift with time they are most useful for strictly time bounded navigation starting from a known state: position, velocity and orientation. This implies that it is necessary to determine the initial conditions using some other positioning method. However, where strap-down inertial technologies really come into their own is when they can be coupled with another navigation system which provides periodic measurements of position, and optionally velocity or orientation. This primary navigation subsystem may be used to continually 'reinitialise' the inertial navigation subsystem so that when it is unavailable for a period of time position outputs are still available thanks to the 'secondary' inertial navigation subsystem.

In a carefully designed hybrid system, the primary navigation subsystem, when it is operating, may be used to calibrate out the major sensor offset errors in the inertial navigation subsystem as well as for providing continually updated 'initial' conditions. This process can help to realise improved performance from low-cost sensors subject to unknown and variable sensor offsets.

Combining different technologies such as inertial sensors with radio positioning systems such as GNSS or others can actually lead to better combined performance than either is able to offer alone. Furthermore the INS provides resilience to short-term loss of radio signals such as when travelling through tunnels or when subjected to radio interference.

There are fundamentally three common methods for combining INS with radiolocation systems such as GNSS:

1. Loosely coupled in which the two systems operate separately but the outputs are compared and the best combined values selected (Figure 5.5). The outputs may optionally be fed back to both for internal corrections.
2. Tightly coupled in which the two are integrated in a single navigation filter (Figure 5.6).
3. Deeply integrated in which the measurements from both systems are combined at the earliest possible stage – even before the navigation filter.

Each level of integration is more complex than the previous, but the additional complexity brings advantages such as, for example, the ability

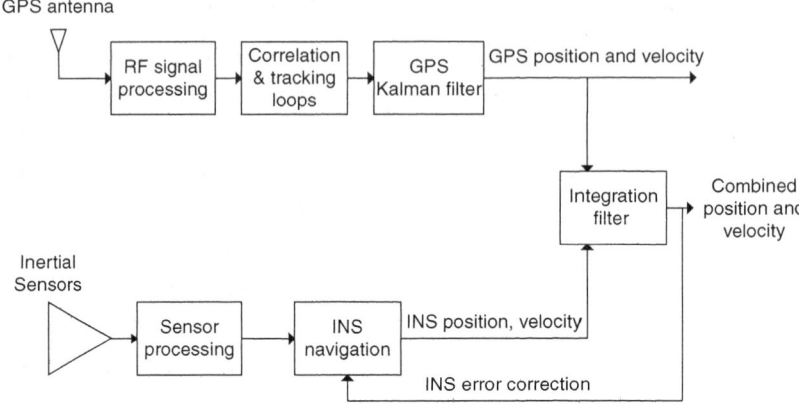

Figure 5.5 Loosely coupled system combining GPS and INS

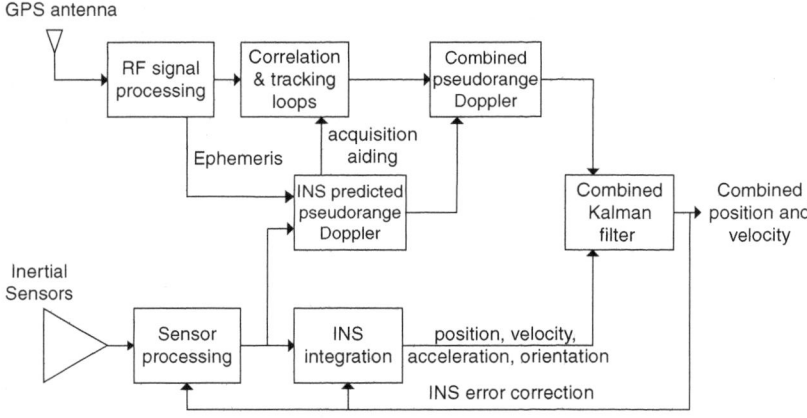

Figure 5.6 Tightly coupled system combining GPS and INS

to utilise measurements from just a single satellite rather than requiring four satellites to compute a navigation solution.

With tighter integration the inertial data provides a reference trajectory which can be used by the GNSS receiver to steer the correlators during signal integration and this may help with weak signal recovery, rejection of jamming signals or faster signal acquisition.

For the INS having regular position measurements from another system, such as GNSS, can enable sensor errors such as bias, scale factor and alignment errors to be calibrated out dynamically and to be tracked as they vary with time. This leads to better performance during periods of unaided navigation using the INS alone. Grewal [32] provides more detail about the integration of strap-down inertial navigation and GPS.

5.7 Dead reckoning for vehicles

In the case of vehicle navigation the situation may be significantly simplified. Vehicles are usually wheeled, which means that they are generally only able to move forwards and backwards in the directions allowed by the wheels, with, typically, one pair of wheels being steerable for turning corners. Vehicles also travel on roads which are substantially level and are subjected to only relatively small gradients and tilts.

With these constraints in mind it is possible to implement a very light weight dead-reckoning system (a subset of the more general inertial navigation problem described earlier) that is able to provide a good estimate of the vehicle's position for short periods of time when GPS navigation is not available. Such simplified systems are widely used in vehicle navigation systems and they typically make use of:

- a single rate gyroscope mounted to measure the turn rate of the vehicle – the rotation around a vertically orientated axis, the vehicle's yaw rate;
- a distance measuring input from the odometer – typically a wheel counter giving a pulse for every unit of distance travelled.

For more sophisticated applications in which the vehicle may not move in the manner described above, where better performance is required or where a wheel counter input is not available, one or two accelerometers may be used to compute the velocity and distance travelled. Alternatively a magnetometer, digital compass, can be used for heading estimation.

5.7.1 Outline formulation of simple loosely coupled vehicle DR system

An Extended Kalman filter (EKF) is typically used to model and track the vehicle position. The vehicle dynamic model is typically formulated around the following state variables:

$$x = \begin{bmatrix} e & n & v & H & a & \omega & \delta SF_d & \delta SF_g & \delta b_g \end{bmatrix} \qquad (5.8)$$

e	easting position in metres
n	northing position in metres
v	velocity in m/s
H	heading (radians)
a	vehicle acceleration
ω	rate of turn (radians per second)
δSF_d	odometer scale factor error
δSF_g	gyroscope scale factor error
δb_g	gyroscope bias

From the GPS receiver we obtain position, velocity (speed over ground) and heading measurements, and from the sensors we get wheel counter pulses and rate of turn. This gives us the following observation variables:

$$z = \begin{bmatrix} \lambda_{gps} & \phi_{gps} & v_{gps} & H_{gps} & N_{odo} & \omega_g \end{bmatrix} \qquad (5.9)$$

λ_{gps}	latitude from GPS
ϕ_{gps}	longitude from GPS
v_{gps}	GPS speed
H_{gps}	GPS heading
N_{odo}	wheel counter pulses
ω_g	turn rate from gyro

From this it is a relatively straightforward task to design the EKF using expected vehicle dynamics and sensor noise parameters to derive the discrete form equations, noting that the GPS receiver typically outputs positions in the WGS84 coordinate frame, whereas we've assumed Cartesian eastings and northings for the vehicle motion.

5.8 Human navigation

In the case of vehicular navigation the use of GPS allows the drift introduced by the sensor offsets to be calibrated out. However, for pedestrian navigation the problem is far more complex.

Firstly we do not have a well-defined model of the motion dynamics. People can move in any direction and are not constrained by wheels and a rigid body. Secondly there is no clearly defined set of body axes to which the sensor platform can be attached. As a result we need to be able to deal with arbitrary alignment of the sensor platform and movement that may be, at times, rather unconstrained.

There are a number of particular approaches that may be used depending on the end application:

- simple pedometer function, step counting, which can be reliably implemented for foot, belt or arm worn devices.
- shoe- or foot-mounted sensors that capitalise on some of the unique benefits that can be obtained during the period when the foot is in contact with the ground;
- head-mounted sensors may be used in certain specialist applications.

In general, integration with GPS is not as useful as in vehicle-mounted applications for several reasons:

- a lot of pedestrian navigation takes place indoors, or in areas with poor GPS coverage;
- the sensor device may suffer from an obscured view of the sky because of the position it is mounted on the body, or because it is underneath clothing or in a pocket or handbag;
- GPS heading and velocity become very unreliable at low speeds, below about 3 m/s, which is the case with most pedestrian applications other than sport.

5.8.1 Determining heading

There are two techniques that can be used to determine heading (orientation):

1. compass using magnetometers;
2. rate gyroscopes.

Both techniques can be used to compute the orientation of the sensor, but knowing the orientation of the person wearing the sensor requires that it be attached to the person in a fixed known orientation. For example, simply placing it in a pocket is unlikely to give a very good indication of the orientation of the person, although it may still be possible to track the orientation of the sensing device.

Using magnetometers as a digital compass can yield reasonable results provided that tilt compensation is included. The algorithms outlined in Section 6.2.2 are for a tilt compensated compass in which an accelerometer is used to measure the gravity vector. The algorithms described can easily be extended to include the inclination (or dip) as an additional output. Without tilt compensation significant errors can be induced. This approach has the advantage that the orientation measured is absolute, and not subject to drift over time, although it is relative to magnetic north and it is also subject to magnetic effects in the environment. In practice a digital compass can give heading accurate to within a few degrees under good conditions.

Using rate gyroscopes, on the other hand, requires a known initial condition (orientation at the start of navigation) or an accompanying positioning technology that can be used to calibrate out the gyroscope bias and scale errors. Using low-cost MEMs sensors drift rates can be quite high, leading to unacceptably large orientation errors after as little as a minute or so of operation.

Ultimately for high-quality personal navigation the combination of magnetometer and rate gyroscopes may yield the best results.

5.8.2 Distance or speed measurement

For human worn applications, the most common measure of distance is to count footsteps. Conventional pedometers use a weighted spring contact that opens and closes with steps taken. It is tuned to resonant frequencies in the typical range of walking to running. Increasingly accelerometers are now used for step counting. Typically they consider only the vertical acceleration component.

Pedometers are usually worn on a waist belt or attached to the shoe, although reliable step counts can also be obtained by shoulder, arm or even head mounted sensors.

For sensors designed to count steps and which can be worn in a specific way (for example the Nike + iPod system) a single axis accelerometer and simple peak detection algorithm can be used. These methods typically apply low-pass filtering with a cut-off around 4 Hz. Step rates are typically in the 1 Hz to 3 Hz range. Since the vertical component also has the gravity vector imposed on it, it is common to use high-pass (0.25 Hz to 0.5 Hz typically) filtered accelerometer data in the step detection algorithm. Since the gravity measurement is affected by the orientation of the accelerometer, this method is sensitive to orientation and changes in orientation. Although peak values can be variable it is relatively easy to extract peaks and classify them as step counts or not based on repetition. Simple algorithms like these have been shown to yield step count error rates of a 'few per cent' for walking and less than 1% error for running.

However, improved performance can be achieved by using a tri-axial accelerometer, especially when the orientation of the sensor device could be variable. In this case the vector norm of each measurement is used as input to the step counting, or behaviour classification, algorithm:

$$A = |a| = \sqrt{a_x^2 + a_y^2 + a_z^2}. \tag{5.10}$$

Computing the norm of each measurement set like this results in a constant gravity offset independent of accelerometer orientation. It can be removed by subtracting the gravity value, rather than resorting to high-pass filtering the data. The resulting measurements may be presented to the simple algorithms described above to extract signal peaks and perform the step counting.

Although very simple algorithms can be used to count steps, it is sometimes useful to use more sophisticated algorithms, perhaps to create a wider-based classification of activity. One technique that is used is to compute a DFT of the sampled accelerometer data in order to extract energy by frequency classification. Typically this is done on an overlapping windowed basis. For example a 64-point DFT (or FFT) might be computed approximately every 1.25 seconds for a 25 Hz sampled accelerometer. Each successive DFT would overlap the previous one by 50%. This gives approximately 0.4 Hz bin resolution and can be used to detect the clear harmonic content introduced by walking. It can also be used to distinguish walking or running from other activities, such as travelling in a vehicle.

Having extracted reliable step count indication, it may be converted into an approximate distance travelled using an average step length constant. However this simple approach introduces large errors, especially when walking around urban environments where crowds, traffic, steps and other obstacles cause a lot of variability in stride length. Whilst there are not yet any perfect solutions, there are a few research areas that show promise:

- In the case of shoe-mounted sensors, there is a period of time when the foot is in contact with the ground and is, therefore, stationary. If one uses this time to measure the accelerometer offset and zero the reading from it, it is possible to compute, using inertial navigation techniques, the precise distance it moves between steps, and thereby arrive at a more accurate estimate of distance travelled.
- Examination of the acceleration profile for a step, and the frequency of steps could be used to infer the particular stride characteristic and adjust the average stride value accordingly. Techniques drawn from machine learning could potentially be used to train the sensor in the walking style of the wearer.

5.8.3 Pressure sensors

Another sensor that is widely proposed for indoor use is an atmospheric pressure sensor (barometer). New miniature sensors capable of 'a few Pascal' pressure resolution (which translates into less than 1 metre height) are able to determine floor level in a multi-storey building. Their measurements vary with weather conditions, but these change relatively slowly and can be removed in the signal processing.

6 Other techniques and hybrid systems

6.1 RFID and RTLS

6.1.1 RFID

RFID (Radio Frequency IDentification) systems are systems in which the identity of a tag is read by a reader or interrogator. This is accomplished when the tag is in the vicinity of the reader's antenna. The reader transmits a radio signal which is picked up by the tag which responds by sending back its identity. Tags may be passive (no battery) or active (containing a battery) and typically range in size from millimetres to centimetres. Passive tags and smaller form-factor tags generally have short reading ranges (centimetres), but active tags may have much longer reading ranges (metres up to several tens of metres). Tags can also be writable allowing the system via the reader to store information in the tag.

Generally RFID tags operate at 13 MHz or in the 840–960 MHz bands (depending on local radio regulations applying in the country of operation). Tags may be very low cost, but readers are relatively high complexity and are correspondingly more expensive.

There are applications in which the use of RFID is perfectly adequate for the purpose of locating or positioning objects. Readers may be placed at strategic 'pinch points' throughout the region of interest. Whenever tags move through the pinch points the reader identifies the tag and a software application keeps track of the zone in which each tag is located. These systems are particularly advantageous when there are very large numbers of tags, but relatively few zones (pinch points).

6.1.2 Real-Time Locating Systems (RTLS)

The RFID industry has been actively embracing the concept of RTLS (Real-Time Locating Systems):

Real-Time Locating Systems (RTLS) seek, continuously and in real-time, to determine the position of a person or object from a distance within a physical space. For example, a physical space can be an entire hospital, a care unit, or a patient's room. More precisely, RTLS concerns people or objects being located within a physical space, without having to move through a portal into or out of that space, and without having to bring an electronic interrogator to the person or object. In this context, real-time means the system checks the location and updates the position of the person or object on a frequent basis and the information it receives concerns the situation at that instant. [33]

RTLS is a 'catch-all' term for a range of different systems and technologies including: UWB (Ultra Wide Band), TOA (Time of Arrival), TDOA (Time Difference of Arrival), AOA (Angle of Arrival), RSSI (Received Signal Strength Indication), infrared, acoustic, ultrasonic, Wi-Fi, ZigBee and RFID.

RTLS is generally understood to exclude navigation systems, including GPS and inertial navigation.

Real-time locating systems may, therefore, be based on many of the technologies described in this book and they can have wide and varied manifestation. As such the reader is directed to the individual chapters and sections for the different technologies.

6.2 Digital compass

6.2.1 The Earth's magnetic field

A compass is an instrument that responds to the Earth's magnetic field, usually by indicating the direction of the strongest field and therefore the north–south field direction. An electronic sensor, called a magnetometer, may be used to measure the strength of the magnetic field surrounding it. With appropriate signal processing the sensor measurements can be used to determine the direction of magnetic north.

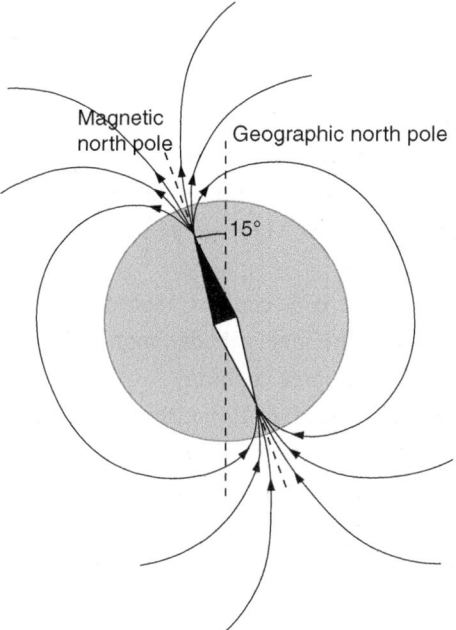

Figure 6.1 Earth's magnetic field

However the magnetic field does not lie horizontal to the surface of the Earth. Figure 6.1 illustrates the formation of the Earth's magnetic field. The field may be considered to be formed around a dipole in the centre of the Earth and aligned to the magnetic north–south poles on an axis approximately 11 degrees tilted relative to the geographic north–south poles (the axis of rotation of the Earth).

This means that at any point on the surface of the Earth the magnetic field has an inclination or dip relative to horizontal. This varies from −90° at the south magnetic pole to +90° at the north magnetic pole. Since the magnetic north pole is not coincident with true north there is also a horizontal angular offset between true and magnetic north pole. This is called the declination and it varies depending on where the observer is.

In addition there are other factors that need to be taken into account if magnetic field measurements are to be used for navigation:

- The location of the magnetic poles moves over time. The amount of drift is, however, small. In 2005 the position of the north pole was estimated to be 82.7°N, 114.4°W, and the south pole 64.7°S, 138.0°E. The north pole is moving by approximately 0.35° northwards and 0.9° westwards every year. The World Magnetic Model is a joint product of the United States' National Geospatial-Intelligence Agency (NGA) and the United Kingdom's Defence Geographic Centre (DGC). The WMM was developed jointly by the National Geophysical Data Center (NGDC, Boulder CO, USA) and the British Geological Survey (BGS, Edinburgh, Scotland).

- The magnetic field is subject to short-term variations, including an underlying diurnal variation in which the magnetic declination in Edinburgh varies by a few tenths of a degree each day.
- It is also affected by local influences. The Hard Iron effect is the effect of magnetic materials making up the instrument used to measure the field. These effects are essentially constant. The Soft Iron effects are caused by factors in the local environment; the presence of magnetic materials in the local environment. They are varying and add noise and uncertainty to the measurements.

6.2.2 Digital compass techniques

The Earth's magnetic field intensity varies from about 24 μT to 66 μT. This field is much smaller than many of the magnetic fields in the environment around us, including the Hard Iron effect.

Given a Cartesian coordinate frame orientated in the usual NED (north east down) configuration, the ideal magnetic sensor would measure field components X, Y, Z leading to the following definitions:

$$
\begin{aligned}
H &= \sqrt{X^2 + Y^2} \\
F &= \sqrt{H^2 + Z^2} \\
I &= \arctan\left(\frac{Z}{H}\right) \\
D &= \arctan\left(\frac{Y}{X}\right)
\end{aligned}
\tag{6.1}
$$

in which H is the horizontal field intensity, F the total field intensity, I the inclination (dip) and D the declination.

What we'd really like to do is to determine the full orientation of the sensor, and one way of doing this is to use the gravity measurement from an accelerometer to electronically 'level' the digital compass.

Given that the defined reference frame, NED, has gravity pointing in the z direction as follows:

$$\begin{bmatrix} 0 \\ 0 \\ g \end{bmatrix} \tag{6.2}$$

we can 'unrotate' the accelerometer measurements solving for the Euler angles of roll and pitch using the following equality:

$$\mathbf{R}_y(-\theta)\mathbf{R}_x(-\phi) \begin{bmatrix} a_x \\ a_y \\ a_z \end{bmatrix} = \begin{bmatrix} 0 \\ 0 \\ g \end{bmatrix} \tag{6.3}$$

which, having substituted the Euler rotation matrices yields a solution for roll:

$$\tan(\phi) = \left(\frac{a_y}{a_z} \right) \tag{6.4}$$

and the solution for pitch:

$$\tan(\theta) = \left(\frac{-a_x}{a_y \sin(\phi) + a_z \cos(\phi)} \right). \tag{6.5}$$

Of course these equations only hold true when the accelerometer is stationary; any linear accelerations will introduce errors.

In order to solve for the magnetic compass bearing we must first subtract the Hard Iron offset from the measurements and then rotate the result through the roll and pitch angles to give us measurements in the NED coordinate frame, allowing the magnetic bearing to be calculated.

Given a vector of magnetometer measurements $\mathbf{B}$ and a Hard Iron offset vector $\mathbf{V}$ and magnetic inclination I, we can formulate the equality:

$$\mathbf{R}_z(\psi) \begin{bmatrix} B\cos{(I)} \\ 0 \\ B\sin{(I)} \end{bmatrix} = \mathbf{R}_y(-\theta)\mathbf{R}_x(-\phi)(\mathbf{B} - \mathbf{V}). \qquad (6.6)$$

Substituting the Euler angle rotation matrices from Section 2.4.3 yields the following equality:

$$\begin{bmatrix} B\cos\psi\cos I \\ -B\sin\psi\cos I \\ B\sin I \end{bmatrix}$$

$$= \begin{bmatrix} (B_x - V_x)\cos\theta + (B_y - V_y)\sin\theta\sin\phi + (B_z - V_z)\sin\theta\cos\phi \\ (B_y - V_y)\cos\phi - (B_z - V_z)\sin\phi \\ -(B_x - V_x)\sin\theta + (B_y - V_y)\cos\theta\sin\phi + (B_z - V_z)\cos\theta\cos\phi \end{bmatrix}.$$

$$(6.7)$$

From this we can easily solve for heading by dividing the x and y components:

$$\tan\psi = \frac{(B_z - V_z)\sin\phi - (B_y - V_y)\cos\phi}{(B_x - V_x)\cos\theta + (B_y - V_y)\sin\theta\sin\phi + (B_z - V_z)\sin\theta\cos\phi}.$$

$$(6.8)$$

Now all that remains is to measure the Hard Iron offset vector. This can be done as a calibration step. Since rotating the sensor through 180° should result in equal and opposite measurements we can simply rotate it looking for the maximum and minimum readings obtainable on each axis – in an environment free of significant Soft Iron effects. The Hard Iron offset vector can be taken as the mean of the minimum and maximum readings obtained for each axis.

This returns a heading, and combined with the accelerometer reading full orientation, in a NED coordinate frame with north aligned to magnetic north. If the sensor is moving and subjected to linear acceleration, the above algorithm needs to be adapted to compensate for the motion. This is left as an exercise for the reader.

6.3 Optical and image tracking techniques

Optical and image tracking techniques have been used for some time in aeronautical and space applications in which the position, and possibly orientation, of an aeroplane, rocket or space vehicle can be determined by processing images captured by at least two geographically separated cameras. Increasingly, with the availability of low-cost digital cameras these techniques are finding their way into mainstream commercial and consumer applications, in particular for computer games and motion analysis.

The principle of operation of most systems falls into one of two architectures that use either 'following cameras', or 'staring cameras'. Beyond this they may rely on infrared light and retroreflective objects or object recognition based on processing the images captured.

6.3.1 Systems using following cameras

In a system based on following cameras one or more cameras, which are mounted in sets of gimbals having a control system that can rotate them around, are pointed at a target (Figure 6.2). The cameras are continually controlled so that the target is kept on the boresight (centre of the image). Measuring the angles from the gimbals allows the direction of the vector along which the camera is pointing to be determined. When two or more cameras are pointing at the same object this allows the 3D position of the

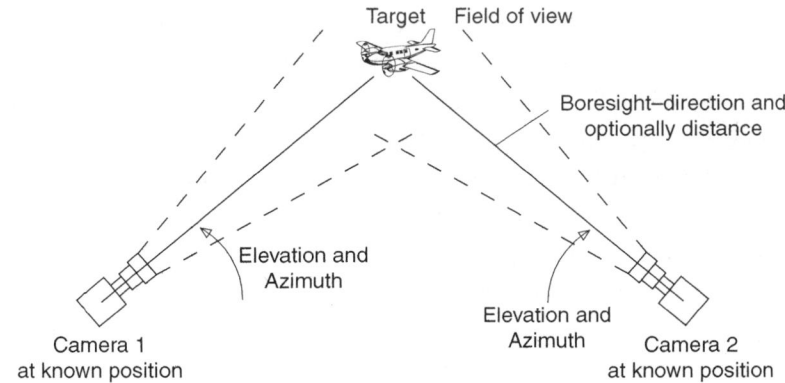

Figure 6.2 Optical tracking using following cameras

object to be determined; assuming, of course, that the cameras were installed at known locations and that the alignment of the gimbals was calibrated.

If distance can be measured then a single following camera is sufficient to determine the target position, otherwise a minimum of two cameras are required. Since the object may become obscured (if it goes behind another object) it is common to use multiple cameras so that at least one (or two) can 'see' the object at all times.

Systems like this are particularly useful in applications for which the direction to the target is the most important attribute, or where there is a need to align a secondary system (such as a microphone, radio antenna or weapon) towards the target. Since the target is aligned with the camera boresight there is no need to calibrate distortion at the edges of lens field-of-view and they can make use of zoom lenses which adapt their focal length appropriately to the situation. The tracking zone is limited by the amount of motion that the gimbals can support which is usually much greater than the field of view of the camera and lens. Their major drawback is the limitation that only one object can be tracked by a camera at any one time.

6.3.2 Systems using fixed cameras

An alternative approach is to use fixed or 'staring' cameras. These are installed at known fixed positions in such a way that two or more cameras can 'see' every point in the desired area of coverage (Figure 6.3). Image processing techniques are used to extract the position of the target(s) within the image of each camera. The displacement from the centre of the image (camera boresight) is measured and given a known focal length (field of view) for the lens the direction of the vector linking the camera to the target can be computed. The intersection of the direction vectors from two or more cameras allows the position of the target to be determined. Usually more than the minimum two cameras are used in order to deal with occlusion of the target and to provide additional redundancy of measurements. It is not uncommon to use six or more cameras.

Systems using fixed cameras can make use of low-cost cameras and standard lenses and it is also possible to track multiple targets simultaneously.

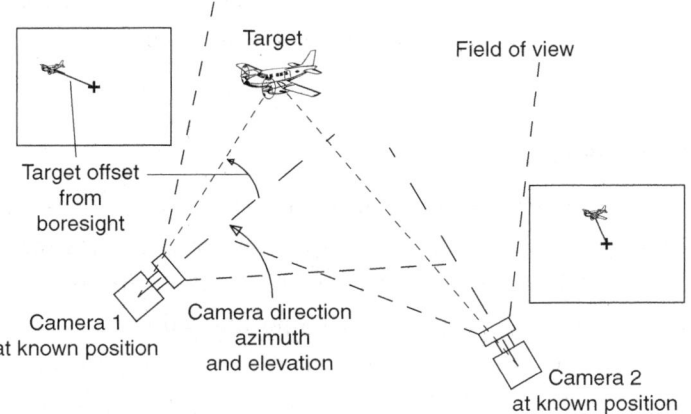

Figure 6.3 System using fixed or 'staring' cameras

An example of this kind of optical tracking system is the Hawk-Eye system from Hawkeye Innovations that is used to track the ball in sports such as cricket, tennis and snooker. It uses at least four, and typically six to ten, cameras arranged around the area. Having identified the precise position of the ball in each image triangulation is used to compute its position accurate to a few millimetres. Sophisticated software that includes an 'understanding' of the nature of the game and the environment in which it is played is used to interpolate the ball's trajectory based on an image frame rate of typically 50 frames per second. This allows the system to very accurately and confidently determine the trajectory of the ball, its speed profile and point of impact with the ground, pads, racket, bat or other objects.

6.3.3 Retroreflective infrared markers

For many applications a similar camera set-up to that described in the previous section is used, except that each camera is also accompanied by an infrared lamp and the object(s) being tracked have small retroreflective patches or balls attached to them. The retroreflective patches appear very bright to the cameras and thus are easy to identify and track. This sort of technology has been used for detailed motion capture in which a number of patches are attached to each limb and joint of a human or animal target in order to measure the way in which it moves, either for research or for motion capture to create computer models of the subject.

6.4 Map matching

For many common transport applications today, such as route guidance and for future applications such as road usage pricing, it is necessary to know exactly where on the road a vehicle is and the direction of travel. These applications usually rely on GPS position information and digital map data, supported by suitable map matching, route planning and guidance software. En-route guidance gives directions to drivers in real-time to assist navigation. In some cases, en-route guidance may be used as the sole means of navigation by users unfamiliar with a particular area (e.g. commonly used in rental cars). In others, it may provide additional information for a user already familiar with an area, allowing them to choose the most appropriate route.

Given knowledge of where the vehicle is, and the driver's required destination, en-route guidance presents navigation directions on a turn-by-turn basis as needed by the driver. The information includes the name of the road on which the vehicle is currently travelling, distance to the next turn, the name of the road to turn into, and the direction of the upcoming turn. These are presented to the driver via visual and audio interfaces. The integration of the vehicle position with digital road maps provides the driver with the majority of the essential information for route guidance (Figure 6.4).

In order to acquire the essential information for route guidance, the vehicle needs to be equipped with a series of interconnected functional subsystems. Each of the subsystems consists of software and hardware components. In general, the subsystems can be categorised as follows:

- Navigation module to determine the vehicle's position and direction based on sensor measurements and GPS position combined with a map matching technique.
- Route guidance to determine the optimal routing between origin and destination or anywhere in-between prior to or during a journey based

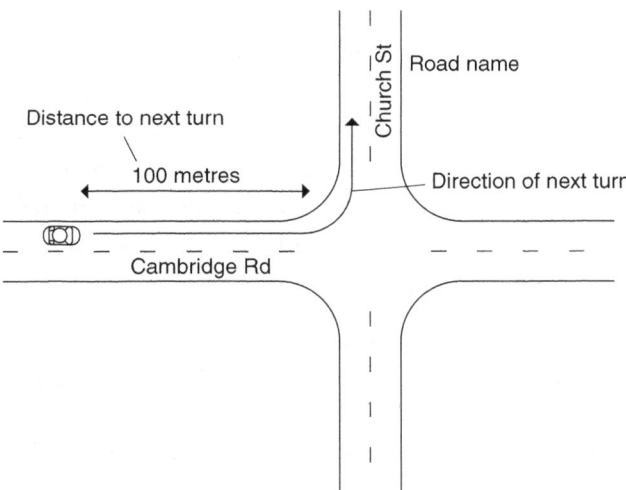

Figure 6.4 Route guidance example

on real-time updates on the state of the road network from the traffic management centre and positioning data from the navigation module.

- Driver support (man–machine interface) which accepts driver inputs and responds with audible or displayed directions as needed.
- Communication module which is used to obtain real-time map updates and news about current traffic conditions.

However, stand-alone GPS suffers from signal masking in areas with heavily tree-lined roads, urban canyons and wooded areas; interference from wireless communications (jamming) as well as signal reflections from buildings, large vehicles and other reflective surfaces (multipath). These lead to poor reliability in some situations, so almost all modern satellite navigation systems make extensive use of map matching. The map matching algorithm is used to integrate the GPS positioning data with the digital map data in order to obtain second-by-second vehicle locations on the road network. Poor map matching can place the vehicle on the wrong road, confusing the driver and making the route guidance ineffective. As a result, the performance of the route guidance, the primary function of the satellite navigation system, largely depends on the performance of the map matching algorithm.

Unfortunately digital maps also have errors. Digital map data is usually based on a 'single line road network' representing the centreline of the road. Road attributes such as width, number of lanes, turn restrictions at junctions, and roadway classification (e.g., one-way or two-way road) often do not exist in the map data. Therefore, the accuracy and uncertainty of digital road network data is a critical issue if the data is used for vehicle navigation. One must be aware of the following concerns regarding the quality of road network data:

- The features (e.g., roundabouts, junctions, medians, curves) of the real-world that have been omitted or simplified in the road map. This is usually known as topological error.
- The accuracy of the classification (e.g., junction or roundabout) of those features.
- Data currency i.e., how recently the map was created.
- The displacement of a map feature (e.g., road centreline, specific junction) from its actual location in the road. This is generally known as geometric error. It may be caused by the digitisation process, or the accuracy of the surveying tools used, but errors are also introduced by discrepancies in the reference datums used. See Chapter 2.

Both geometric and topological errors of map data may introduce significant horizontal errors in land vehicle positioning and navigation. Although the quality and accuracy of digital road map data is improving rapidly, errors are still typically in the range of 2 to 10 metres.

According to Quddus [34] no less than 35 new map matching algorithms were produced and published in the literature between 1989 and 2006. This is largely fuelled by the growth in demand for Intelligent Transport Systems (ITS). The algorithms used can be broadly classified into four categories of algorithms: geometric; topological; probabilistic and other advanced techniques.

6.4.1 Geometric

A geometric map matching algorithm makes use of the geometric information of the spatial road network data by considering only the shape of the links and not the way links are connected to each other. The most commonly used algorithm is a simple search algorithm. In this approach, each of the positioning fixes is matched to the closest 'node' or 'shape point' of a road segment.

Extensions to geometric algorithms incorporate matching to the nearest point on segments connecting the nodes, and they may also take into consideration the trajectory of the vehicle. Geometric algorithms have relatively low complexity which comes with limitations in performance.

6.4.2 Topological

Map matching algorithms which make use of the geometry of the links as well as the connectivity and contiguity of the links are known as topological map matching algorithms.

Generally these algorithms are based on correlation between the trajectory of the vehicle and the topological features of the road (turns, curvature and connections). A number of conditional tests are applied to eliminate road segments that do not fulfil some pre-defined thresholds. Several variants of these algorithms have been proposed with particular emphasis on selecting the conditional tests and reducing computational complexity. The algorithms generally perform better than simple geometric methods, but struggle with junctions and when vehicle speed is low because the GPS trajectory becomes noisy and error prone at low speeds.

6.4.3 Probabilistic

Probabilistic algorithms require a confidence region to be generated by the navigation sensor for each position fix. These are usually elliptical and define the region within which the position fix is likely to lie. The error region is then superimposed on the road network to identify a road segment on which the vehicle is travelling. If an error region contains a number of segments, then the evaluation of candidate segments is carried out using heading, connectivity and closeness criteria.

6.4.4 Other advanced map matching techniques

Advanced map matching algorithms is where most of the current research is focussed and there are many approaches being tested including:

- Kalman Filters and Extended Kalman Filters
- Dempster–Shafer's mathematical theory of evidence
- Flexible state-space models
- Particle filters
- Fuzzy logic models
- Bayesian inference

Descriptions of these different techniques are beyond the scope of this book and the reader is referred to the literature for further information. Quddus [34] is a good starting point for further research.

6.5 Image recognition

Although still in the early stages of maturity image recognition is likely to become a very important element in location and positioning systems of the future; after all this is one of the primary methods we, as humans, use to locate ourselves and navigate the world around us. At present image recognition is limited to relatively simple tasks such as: identifying iconic landmarks; or reading specific features such as street signs. However, image processing techniques are one of the fundamental methods used in many simultaneous localisation and mapping (SLAM) systems [35]. See Section 6.7 for a more detailed introduction to SLAM.

6.6 Fingerprinting

In systems using Wi-Fi signals for localisation the so-called 'fingerprint-ing' technique is most commonly used. The environment within which localisation is required is characterised by making a comprehensive set of signal strength measurements of all APs (Access Points) within range across the entire region of interest. These measurements are stored as a set of fingerprints. When a device moves to an unknown position, the measured signals are compared to the fingerprint measurements and the most likely position identified based on the match obtained.

Such systems are widely claimed to be accurate to about 5 metres. However, performance is very much determined by the environment and the way in which the system is used. In general it relies on having sufficient APs receivable throughout the area and for the environment to be relatively static and unchanging. It is also dependent on the time-consuming task of making all the fingerprint measurements and constructing the signal database.

In general it performs much better indoors than outdoors. The walls and building construction lead to feature rich fingerprints that enable good room-level localisation to be achieved. Outdoors, RSSI is a poor indicator of range, being very sensitive to local obstructions and particularly unreliable for body-worn sensors. Retscher [36] shows that Wi-Fi systems can achieve better than 4 m indoors, but outdoor accuracy is more generally 10 m to 40 m.

It has also been proposed to use other sensor measurements, including sound and light (colour and brightness) to construct additional fingerprint characteristics for the operating environment. Research has shown that these additional sensor measurements can help, but more research is needed before they become serious players amongst location and positioning technologies.

However, fingerprinting techniques come into their own when combined with other localisation technologies, and particularly when used in a SLAM configuration as described in the next section.

6.7 Simultaneous localisation and mapping (SLAM)

SLAM (Simultaneous Localisation and Mapping) is a technique arising from the field of robotics, in which the problem is to establish whether it is possible for a robot placed at an unknown position in an unknown environment to build a map of the area and simultaneously to determine its position within the environment. Although arising from the field of robotics, the techniques employed for SLAM may be equally applicable in more general indoor positioning applications.

The 'solution' of the SLAM problem has been one of the notable successes of the robotics community over the past decade. SLAM has been formulated and solved as a theoretical problem in a number of different forms and it has also been implemented in a number of different domains from indoor robots to outdoor, underwater and airborne systems. At a theoretical and conceptual level, SLAM may be considered a solved problem. However, a lot of work remains to practically incorporate SLAM systems into everyday location and positioning applications and products.

6.7.1 Principles of SLAM

SLAM is a process by which a mobile device can build a map of the environment around it and at the same time use this map to estimate its location. Both the trajectory of the device and the locations of all landmarks are estimated without the need for any a-priori knowledge of locations.

A sensor on the device is used to make observations of landmarks within the environment. Optical imaging sensors have been the basis of much of the research to date, but other sensors can be used too: for example sensing radio signal sources such as Wi-Fi Access Points. The observation may include an estimate of the angle (or bearing) to the landmark, or an estimate of the range to the landmark or both angle and range.

It is also necessary to measure the trajectory of the device, and this is usually done using dead-reckoning sensors such as odometer and heading (from magnetometer or gyroscope), or using a strap-down inertial navigation platform.

As the device moves through the environment making relative measurements of landmarks (Figure 6.5), the following data is accumulated:

- a state vector defining the position and orientation of the device within the environment, including a history of past positions;
- observations of the landmarks taken over time;
- estimated positions of the landmarks.

The solution is a probabilistic one of solving the joint probability distribution function of the mobile device and landmark positions as it

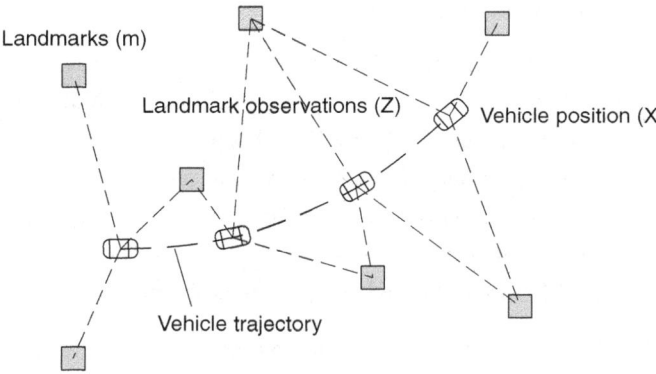

Figure 6.5 SLAM principle

moves around the environment. Since a large part of the error is caused by the unknown position of the mobile device, it will be appreciated that observations of landmarks are significantly correlated with their relative positions being more certain than their absolute positions. Therefore as more observations are gathered from different positions while the mobile device moves around the environment the solution converges.

There are three main approaches to SLAM:

- The most common is in the form of a state-space model with additive Gaussian noise leading to the use of the extended Kalman filter (EKF).
- An important alternative is to describe the vehicle motion model as a set of samples of a more general non-Gaussian probability distribution. This leads to the use of the Rao–Blackwellised particle filter, or FastSLAM algorithm.
- A third approach that is emerging is the use of Gaussian Processes and techniques drawn from the field of machine learning. See Rasmussen [37] for more information.

Note that the basic SLAM problem is formulated for one mobile device moving around a static environment in which the landmarks are stationary. Whilst this is not essential, extensions to non-static environments and multiple mobile devices are non-trivial and are still the subject of ongoing research.

6.7.2 Extended Kalman filter approach

In this approach an extended Kalman filter model is constructed in which state variables represent the motion trajectory of the mobile device and the locations of all the landmarks. The EKF has been widely used in navigation applications, and its primary limitations are well understood, in particular location for navigation is a highly non-linear problem, and the EKF is based on linearised models involving compromises which can lead to unexpected behaviour. SLAM adds additional complexity to the problem:

- The fact that all landmarks need to be updated with each observation leads to a computational requirement that grows quadratically with the number of landmarks $O(N^2)$. However, ongoing research is leading to techniques for reducing the computational load.
- EKF solutions are fragile when poor landmark observations (data outliers) occur. Once the measurements have been incorporated in the covariance matrix, they are hard to undo.

6.7.3 Particle filter approaches

Because of the high-dimensional state space (many unknown landmark positions) of SLAM it is infeasible to directly apply particle filters to the entire problem – the computational load is simply far too high. However it is possible to reduce the state space by applying Rao–Blackwellisation (R–B) in which the state space is refactored as the product of independent distributions. In this case the SLAM problem is factored into a vehicle component and a map component for the landmarks.

The vehicle component is presented as a distribution on the trajectory, rather than a single point, because this leads to the landmark points being independent. The landmark positions are represented as independent Gaussians. Further detail of this approach may be found in Sim [35].

6.7.4 Using Wi-Fi measurements with SLAM

As described earlier the use of signal strength measurements is a very imprecise way of determining range and thence position, due to variability of propagation. The fingerprinting methods described earlier rely on fully characterising the radio environment and building models of expected signal strengths throughout the area of interest as part of the initial installation and set-up for systems based on the method. This is a major drawback, being costly, time consuming and unable to adapt to changes in the environment.

Ferris [38] describes a novel technique for building wireless signal strength maps without requiring any location labels in the training data. It uses a Gaussian processes latent variable model (GP-LVM) for mapping high-dimensional data to a low-dimensional space. In this way the mobile device is able to build its own spatial representation of the environment based on sequences of raw unlabelled signal strength data, and from this to determine its location.

7   Techniques and performance

7.1 Understanding accuracy and precision

Determining the position of an object is a statistical task. It is based on measurements and observations of the environment around and signals from neighbouring devices – which are also affected by the environment. All of the measurements are subject to noise and uncertainty. When the measurements are combined to compute an estimate of the position, the result is statistical. This means that any position we compute has a probability and error margin associated with it.

It is common practice to describe a positioning system as 'accurate to 2.5 metres' (or some other number according to the system and technology used). So what does this actually mean?

What it *does not* mean is that every single location it ever generates will be within 2.5 metres of the 'actual position'! So this raises several important questions:

1. What proportion of measurements are within 2.5 metres?
2. How do we know when a measurement is worse than the specified value?
3. Under what conditions would we expect this performance to be achieved?
4. What is the 'actual position' with which the system is being compared?
5. Are the errors repeatable? In other words will the same error occur if the test position is revisited at a later time?

Understanding, or at least an awareness, of these factors is essential for anyone who wishes to deploy real-world location-based services and applications. Most vendors do not answer all, and often answer none, of the questions raised above. The answers are usually unsaid or implied, or even worse unknown, and it is very much a case of 'buyer beware'.

7.1.1 Accuracy and precision

The term 'accuracy' is almost universally used to describe locating and positioning systems, although very often what is actually meant is 'precision', also sometimes called 'repeatability'. However, we should take care that some vendors use the term 'precision' where we'd use the term 'resolution'.

Accuracy and precision differ in their use of 'ground truth' or 'actual position'.

In this book we define these various terms as follows:

'Ground truth' is the actual position using the defined coordinate system within which the system is operating. It is important to be consistent and precise with the coordinate system being used, because as was shown in Chapter 2 transforming between different coordinate systems is not always trivial.

'Accuracy' is a statistical measure of the deviation between the estimated position and the actual position or ground truth for the device. For example a CEP (Circular Error of Probability) of 2.5 m would state that 50% of the estimated positions lie within 2.5 m of the ground truth or actual position of the device, both described in the chosen coordinate system.

'Precision' or **'repeatability'** is a statistical measure of the distribution or scatter of estimated positions without reference to a known ground truth. For example a CEP of 2.5 m would state that 50% of estimated positions lie within a 2.5 m radius circle, but it does not state where the centre of the circle is located.

'Resolution' is used to define the numerical resolution of the position generated. So the system might output the estimated position to 1 millimetre resolution and yet have a specified accuracy of 2.5 m. So giving the position as $x = 25.432$, $y = -4.789$ in metres, has 1 mm resolution, but the accuracy is still 2.5 m.

The difference between accuracy and precision (repeatability) is illustrated in Figure 7.1:

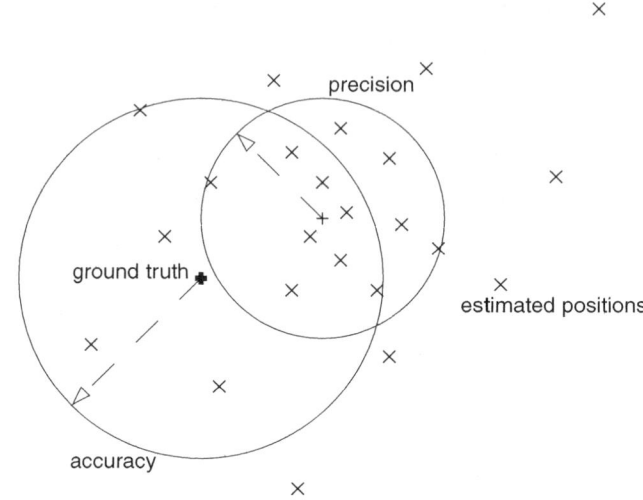

Figure 7.1 Illustration of accuracy versus precision

7.1.2 CEP

A simple accuracy statement, such as 'accurate to 2.5 metres', would normally be a basic circular error of probability (CEP) statement meaning that 50% (half) of the positions reported would lie within a circle of 2.5 m radius on a plane – i.e. two-dimensional positioning. Figure 7.1 is an illustration using CEP. If a 3D position is being described this would normally be explicitly specified. The statement says nothing about the test conditions, nor about the 50% of measurements that lie outside the circle. It also does not explicitly say that the centre of the circle lies at the actual position or ground truth, and therefore even though the statement uses the word accuracy, it may actually mean precision.

Basic CEP is the most common measure of 'accuracy' used for positioning and location systems. Sometimes it is stated for a different confidence level, such as 'CEP 95' in which case it means that 95% of reported positions lie within a circle having a radius equal to the stated error.

It is also worth noting that the performance statistic quoted is often drawn only from those positions reported. Therefore if one took a device to a particular location and tried to measure its position 100 times but only 50 were successful, the accuracy achieved might be quoted as the CEP formed by 50% of the 50 successful measurements, and those 50 points at which a position could not be computed would be discarded from the computation. Whilst only a minority of vendors produce misleading performance figures like this it does happen and it is important for anyone deploying real location-based services and applications to be aware of it.

7.1.3 Statistical measures of accuracy

Since the 50% performance level quoted by many vendors is far too low to be useful in most real-world applications, an alternative approach is to assume that the distribution of positions is Gaussian and compute the standard deviation and based on this to quote the 67% and 95% figures,[1] these corresponding to DRMS (distance RMS) and 2DRMS (two-sigma distance RMS).

In order to compute the DRMS and 2DRMS figures we are going to assume that the horizontal (x) and vertical (y) errors are normally distributed. Whilst not strictly true this is the assumption used as the basis for most error statistics used in the industry. Together they comprise a bivariate (2D) normal (Gaussian) distribution centred at μ_x, μ_y. The means and standard deviations are given by:

$$\mu_x = \frac{1}{n}\sum_{i=1}^{n} x_i$$
$$\mu_y = \frac{1}{n}\sum_{i=1}^{n} y_i \tag{7.1}$$

---

[1] Several different figures ranging between 65% and 68% and 95% to 98% are used in industry as corresponding to DRMS and 2DRMS. This arises because the error distribution is elliptical rather than circular.

$$\sigma_x = \sqrt{\frac{\sum\limits_{i=1}^{n} (x_i - \mu_x)^2}{n}}$$

$$\sigma_y = \sqrt{\frac{\sum\limits_{i=1}^{n} (y_i - \mu_y)^2}{n}}$$

$$(7.2)$$

These are the root mean square (RMS) values of the error. To obtain the DRMS and 2DRMS values we combine the individual axis RMS values as follows:

$$DRMS = \sqrt{\sigma_x^2 + \sigma_y^2}$$
$$2DRMS = 2\sqrt{\sigma_x^2 + \sigma_y^2}$$

$$(7.3)$$

Note that it is based around the computed mean which is the centre of the 'precision' circle indicated in Figure 7.1 and therefore gives a measure of precision and not accuracy relative to ground truth.

Whilst DRMS (and CEP) is a useful method for describing 'accuracy' using a single measure, it is usually only an approximation. Firstly the error distribution is seldom circular, and is better described as an error ellipse. Secondly positioning errors are seldom Gaussian and therefore Gaussian statistics are sometimes a poor way of describing the error distribution.

7.1.4 Error ellipses

Position errors are seldom circular, and the error distribution is better described as an ellipse. This is particularly important for applications in which there might be poor geometries used and also when one has a chain of measurements based on each other in which it is necessary to propagate the total error across them all.

In order to derive the shape and orientation of the error ellipse we first compute the covariance matrix as:

$$\mathbf{P} = \begin{bmatrix} \sigma_x^2 & \sum_{i=1}^{n} \frac{(x_i-\mu_x)(y_i-\mu_y)}{n-1} \\ \sum_{i=1}^{n} \frac{(x_i-\mu_x)(y_i-\mu_y)}{n-1} & \sigma_y^2 \end{bmatrix}. \tag{7.4}$$

The covariance matrix has special properties. Of particular interest to us is that there exists a matrix being a rotated version of our covariance matrix that is diagonal – i.e. the off-axis components of the matrix are zero. The two diagonal elements of this rotated matrix are called the Eigenvalues. The eigenvalues are the square of the magnitude of the major and minor axes of the error ellipse. The sum of the two eigenvalues is equal to the sum of the two variances used to compute the DRMS value in the previous section.

The Eigenvectors of the covariance matrix define the rotation and give the directions of the major and minor axes of the ellipse. By definition the vector $\mathbf{x}$ is an eigenvector of the matrix $\mathbf{A}$ with eigenvalue λ if the following holds true:

$$\mathbf{Ax} = \lambda\mathbf{x}. \tag{7.5}$$

The reader is referred to any of the good books of mathematics for more information about eigenvectors and eigenvalues. For the purposes of this discussion we will skip the detailed derivation which leads to a straightforward solution for the error ellipse in two dimensions:

$$\theta = \frac{1}{2} \tan^{-1}\left(\frac{2\sigma_{12}}{\sigma_{22} - \sigma_{11}}\right)$$

$$a^2 = \frac{1}{2}\left[\sigma_{11} + \sigma_{22} - \sqrt{(\sigma_{22} - \sigma_{11})^2 + 4\sigma_{12}^2}\right]. \tag{7.6}$$

$$b^2 = \frac{1}{2}\left[\sigma_{11} + \sigma_{22} + \sqrt{(\sigma_{22} - \sigma_{11})^2 + 4\sigma_{12}^2}\right]$$

In which θ is the rotation angle of the ellipse and a and b are the minor and major axes, so $a/2$ and $b/2$ are the smallest and largest radial errors corresponding to one standard deviation. The ellipse encloses the area containing 63% (approximately) of the data points.

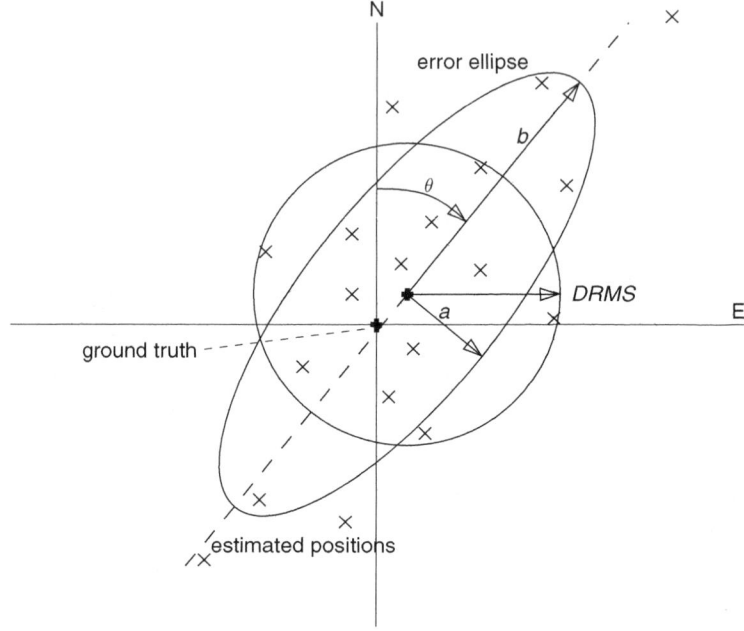

Figure 7.2 Error ellipse

Figure 7.2 illustrates the principles of the error ellipse.

7.1.5 Statistical distribution of position errors

The previous discussions about errors have assumed that measurement errors are Normally distributed (Gaussian distribution). The first question is: Are they? Secondly how do measurement errors translate into uncertainty of the calculated position?

7.1.5.1 Position error distribution

Suppose for now that the measured positions are represented in a Cartesian coordinate frame in which the x and y errors are independent and are Normally distributed relative to the mean calculated position. The position error is given by the square of the sum of the two components:

$$d_i = \sqrt{(x_i - \bar{x})^2 + (y_i - \bar{y})^2}. \tag{7.7}$$

If the x and y positions are normally distributed, as has been assumed, the distribution of the errors d which results follows the Rayleigh distribution, which has the following probability density function:

$$f(x, \sigma) = \frac{x}{\sigma^2} e^{\frac{-x^2}{2\sigma^2}}, x \geq 0, \sigma > 0 \tag{7.8}$$

and the cumulative distribution function:

$$F(x) = 1 - e^{\frac{-x^2}{2\sigma^2}}, x \geq 0, \sigma > 0. \tag{7.9}$$

These are shown graphically in Figure 7.3 for values of $\sigma = 0.6, 1.0, 2.0, 4.0$.

This shows the interesting, and obvious, fact that the most likely size of the position error is not zero, and also illustrates that the maximum error is, in theory, unbounded.

7.1.5.2 Distribution of x and y positions

Most radio positioning systems rely on some form of measurement of the time of arrival of a radio signal at a receiver. Angle measuring and signal strength based systems being exceptions. There are two main components in time-of-arrival measurement errors: (a) noise in the measurement process; and (b) errors introduced by the signal propagation.

It is not unreasonable to model noise in the measurement process as Gaussian, but errors in signal propagation are seldom Gaussian. Refer back to Section 4.6.2 discussing channel models to see that the received signal comprises the sum of a direct path signal and a number of delayed reflected signals.

Radio engineers have developed sophisticated channel models for radio propagation. However, the channel tends to be modelled statistically using either a Rayleigh or Ricean distribution. Both of these are skewed distributions with a longer tail in one direction. The Rayleigh

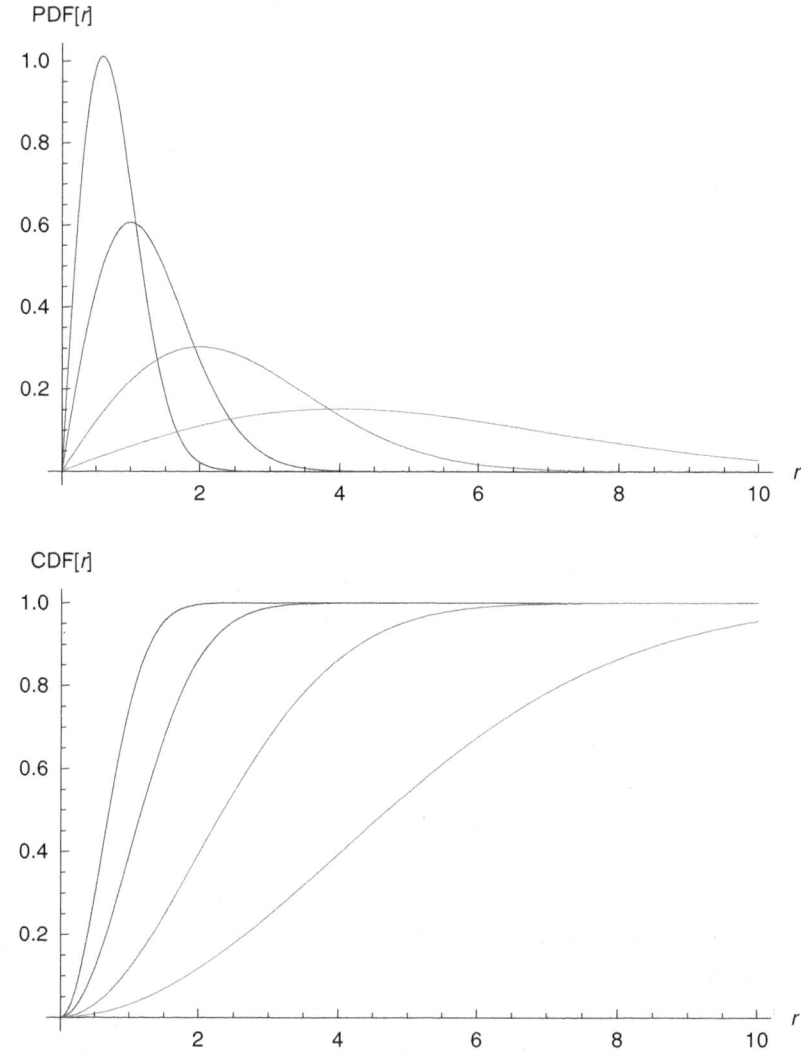

Figure 7.3 PDF and CDF for Rayleigh function, $r = 0.6, 1, 2, 4$

distribution generally provides a better fit for channels in a heavily built up urban environment in which there is no dominant signal along a direct line-of-sight path. It is also considered to be a good fit for tropospheric and ionospheric propagation. The Ricean model is a better fit for channels in which there is a dominant line of sight.

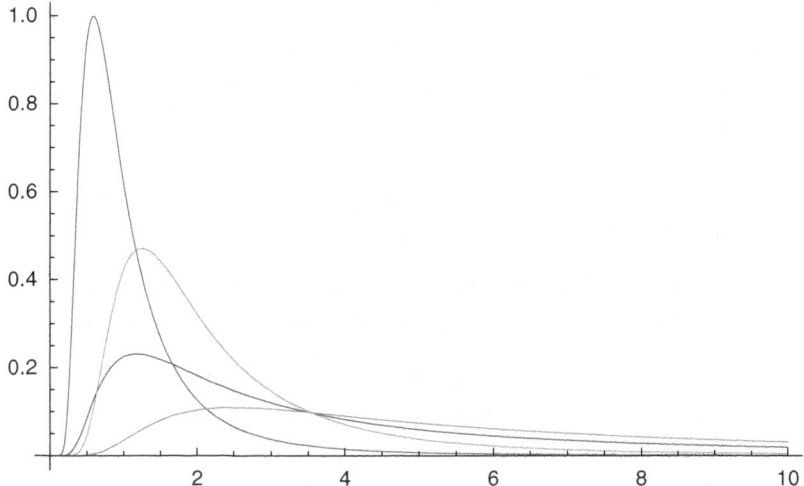

Figure 7.4 Inverse Gamma distribution examples

For propagation in a nearly homogenous environment (radio signals in the atmosphere) the direct path is the earliest arriving signal and the reflected signals are delayed relative to it. For the purposes of positioning we are always striving to measure the time of the earliest arriving signal, but this is not always possible and in heavily multipathed environments the direct path may not be present at all, or may be obscured by later arriving signals.

Therefore it is a poor assumption that the measured ranges have errors that are Normally distributed, and from this it follows that it is a poor assumption that the x and y positions that are computed as a result are Normally distributed. From this it follows that a Rayleigh distribution is at best an approximation to the distribution of positioning errors. Tennina [39] presents the results of research in which they show that the family of Pearson distributions provides a good fit for the statistical distribution of positioning errors in radiolocation systems.

Of particular interest is the Pearson V distribution which is also referred to as the Inverse Gamma distribution. In general the real distribution is more skewed than the Rayleigh with a flatter more extended tail, as illustrated in Figure 7.4 plotted for two different shape and two different scale parameters.

It is beyond the scope of this book to go into more detail about the statistics of positioning errors, especially as the measurement techniques used and methods for estimating the position also need to be taken into account.

7.1.6 3D position errors

The discussion in the previous section was in the context of planar (two-dimensional) positioning. The same principles apply in three dimensions:

- The CEP becomes an SEP (Spherical Error of Probability) and the Distance RMS values become spherical radial measures rather than circular radial.
- The elliptical error distribution becomes ellipsoidal with three axes and two angles for orientation.

In practice SEP and other three-dimensional error statistics are rarely used because most practical radiolocation systems, including GPS, have geometries that are generally planar or one sided, which normally means that the system is less accurate in terms of height than (x,y). Therefore, very often the planar accuracy and height accuracy are separately quoted by manufacturers.

7.1.7 Dilution of precision

The term Dilution of Precision (DOP) is used to describe the way in which a set of measurements will affect the estimated position of the device. DOP is a measure of the 'geometry' of the solution extracted from the relative positions of the device and its neighbours that are used to compute the position. Consider Figure 7.5:

The example on the left shows a good geometry with tight intersection between the measurements from A and B. The one on the right shows a poor geometry in which the intersection between the measurements is elongated. The one on the left results in a nearly circular error probability, whereas the one on the right leads to a more elliptical error probability which is also much bigger than the one on the left.

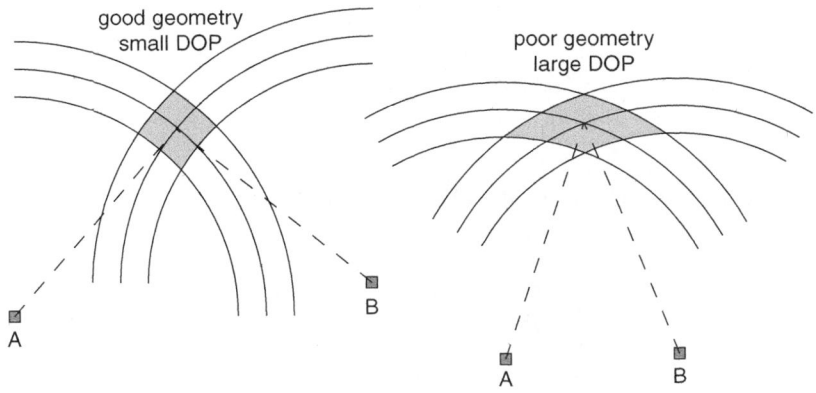

Figure 7.5 Illustration of good and bad geometry

DOP is a simple measure used to describe this effect. It is represented as a unitless number which acts as a multiplier to the anticipated error. Therefore if we compute a position using measurements having a DOP of 2, we would expect the resulting error to be statistically twice as large as a position computed using the same system under similar conditions using measurements having a DOP of 1.

The term DOP has become widely used in GPS systems although it is equally usable in other radiolocation systems. The following illustrates the principle of computing DOP for a GPS measurement, omitting the mathematical derivation and details.

Construct unit vectors to each satellite as follows:

$$\left[\frac{x_i - x}{R_i} \quad \frac{y_i - y}{R_i} \quad \frac{z_i - z}{R_i} \right] \tag{7.10}$$

where R_i is given by:

$$R_i = \sqrt{(x_i - x)^2 + (y_i - y)^2 + (z_i - z)^2} \tag{7.11}$$

with the receiver at position (x,y,z) and the ith satellite at position (x_i,y_i,z_i).

Given four satellites (in order to solve for x, y, z and t) we construct the matrix **A**:

$$\mathbf{A} = \begin{bmatrix} \frac{x_1-x}{R_1} & \frac{y_1-y}{R_1} & \frac{z_1-z}{R_1} & -1 \\ \frac{x_2-x}{R_2} & \frac{y_2-y}{R_2} & \frac{z_2-z}{R_2} & -1 \\ \frac{x_3-x}{R_3} & \frac{y_3-y}{R_3} & \frac{z_3-z}{R_3} & -1 \\ \frac{x_4-x}{R_4} & \frac{y_4-y}{R_4} & \frac{z_4-z}{R_4} & -1 \end{bmatrix} \tag{7.12}$$

from which we can derive the **Q** matrix as follows:

$$\mathbf{Q} = \left(\mathbf{A}^T\mathbf{A}\right)^{-1} = \begin{bmatrix} \sigma_{xx} & \sigma_{xy} & \sigma_{xz} & \sigma_{xt} \\ \sigma_{yx} & \sigma_{yy} & \sigma_{yz} & \sigma_{yt} \\ \sigma_{zx} & \sigma_{zy} & \sigma_{zz} & \sigma_{zt} \\ \sigma_{tx} & \sigma_{ty} & \sigma_{tz} & \sigma_{tt} \end{bmatrix}. \tag{7.13}$$

From this matrix we compute the various measures of DOP as follows:

$$\begin{aligned} PDOP &= \sqrt{\sigma_{xx}^2 + \sigma_{yy}^2} \\ HDOP &= \sigma_{zz} \\ TDOP &= \sigma_{tt} \\ GDOP &= \sqrt{PDOP^2 + HDOP^2 + TDOP^2} \end{aligned} \tag{7.14}$$

where PDOP is the two-dimensional position DOP, HDOP the height DOP, TDOP the time DOP and GDOP the total geometric DOP.

7.2 Theoretical bounds for accuracy

The process of measuring a radio signal and using the measurement to calculate the position is statistical in nature because the signal and measurement are affected by noise (errors in the measurement). Therefore the task is to use noisy measurements of a noisy signal to make the best possible estimate of its position.

A technique called Maximum Likelihood (ML) has been developed within the branch of statistics called Estimation Theory. Named after the mathematicians Harald Cramér and Calyampudi Radhakrishna Rao, the Cramér–Rao lower Bound (CRLB) is the lower limit to variance that can be attained by an unbiased estimator of a parameter θ of a distribution.

An estimator θ^* of θ is said to be 'unbiased' if its expectation is equal to θ for all values of θ:

$$E[\theta^*] = \theta. \tag{7.15}$$

The CRLB states, in its simplest form, that the variance of any unbiased estimator is at least as high as the inverse of the Fisher Information. In mathematical statistics and information theory, the Fisher Information is the variance of the score.

It is possible to derive the Fisher Information and hence the CRLB for the measurements used in a positioning system, and to arrive at a lower bound for the system accuracy. This lower bound describes the best performance that can be achieved for an efficient minimum variance unbiased estimator. Some practical systems today come close to theoretical performance, but most fall short by significant margins.

Apart from the difficulty of constructing an optimum estimator, the measurements used in most location systems are not Gaussian and are often not linear either.

For a system based on wideband signals in which the received signal is cross-correlated with the expected (transmitted) signal it has been shown that the standard deviation of the ranging error (in seconds) follows the relationship:

$$\sigma \propto \frac{1}{\sqrt{\left(\frac{S}{N}\right)WT}} \tag{7.16}$$

where S/N is the signal-to-noise ratio, W is the signal bandwidth and T the integration time.

The constant of proportionality needs to be derived for the specific implementation of the system in question in order to define the actual numerical values for the lower bound. This has been done for GPS arriving at a proportionality constant of $3.444.10^{-4}$ when the C/A code is used for range measurement. Weill [40] provides an insight into the theoretical performance that can be achieved with GPS.

In practice this means that the lower bound for a satellite range measurement with C/N of 35 dBHz and 20 ms integration time (one data symbol and 25 Hz response frequency) is 4.2 m. Obviously, a better result is achieved with improved C/N and longer integration time.

The CRLB has also been derived for UWB systems in which a very short pulse of energy with extremely wide effective bandwidth is used to measure the observed time of arrival of the signal. In this case the bound for range error variance (in seconds) is given as:

$$\sigma = \frac{1}{\sqrt{8\pi^2 B^2 \left(\frac{S}{N}\right)}} \tag{7.17}$$

where B is the 'effective bandwidth', inversely related to the pulse width, and S/N is the signal-to-noise ratio.

Similarly it can be shown for a narrowband phase measuring system the relationship of the range error variance (in seconds) is as follows:

$$\sigma = \frac{1}{2\pi f_c \sqrt{2\left(\frac{S}{N}\right)T}} \tag{7.18}$$

where f_c is the carrier frequency, S/N is the carrier signal-to-noise ratio and T is the integration time (approximately the inverse of the bandwidth of the carrier tracking loop), and in this case the result is equal to the lower bound. Tao Jai [41] presents the derivation of a CRLB for TOA-based Localisation, and Reece [42] shows tighter alternatives to the CRLB for discrete-time filtering.

7.3 Techniques for assessing quality of a position fix

GPS receivers typically report a few parameters broadly related to fix quality: number of satellites, signal-to-noise ratios and DOP. However, they seldom report true fix quality expressed as an estimated positioning error, or the parameters or the error ellipse. Most terrestrial positioning systems are just as remiss, but there are some systems available which are able to report more detailed quality information to the user application.

DOP and the parameters of the error ellipsoid as described earlier in this chapter are relatively easy for the positioning system to compute;

obtaining a reliable estimate of the accuracy is much harder to do. However, there are a number of things the system can do in an attempt to estimate a trustworthy location accuracy figure:

- Firstly the underlying performance of the system under good conditions needs to be understood;
- Next it needs to be able to find or estimate other physical system errors such as uncertainty in the position of satellites or terrestrial anchors (access points or readers);
- It also needs to know or estimate clock and timing errors in the system, satellites, anchors and other components of the system;
- Signal-to-noise ratio may be taken into account, particularly at low S/N ratios;
- Information about the quality of the time-of-arrival estimate for the signal can often be found by looking at noise in the correlator, quality of the signal decoder eye, or phase noise on the received carrier;
- The presence of adjacent channel noise might indicate possible interference;
- Wide temporal fluctuations in signal strength of received signals may indicate the presence of multipath or other non-line-of-sight effects, in the case of GPS wide disparity between different satellite signal strengths could also be indicative of multipath;
- Rapid fluctuations in individual pseudo-ranges beyond the normal dynamic bandwidth of the object being located could indicate multipath or non-line-of-sight effects;
- Having computed the position, looking at the range residuals may indicate general or specific problems in the position solution.

Using these and other measurements and observations coupled with smart algorithms can enable usable estimates of the location fix accuracy to be obtained. The specific algorithms used to combine available quality indicators are dependent on the particular implementation, so it is left as an exercise for the reader to investigate further. Since this information is extremely valuable to application developers, vendors of positioning and locating systems should be encouraged to incorporate reliable estimates of fix quality into their products.

7.4 Tracking the motion of the target

In most systems the output positions are filtered using a Kalman filter, or in some cases a particle filter, to generate the final reported location taking into account the dynamic behaviour characteristics of the tracked object.

The Kalman filter is an optimal estimator. It is used to infer parameters of interest from indirect, uncertain or noisy observations or measurements. It is recursive so measurements may be processed as they are made. It is a linear estimator which is optimal for Gaussian noise.

Even though location and positioning systems are highly non-linear, the Kalman filter has firmly established itself as the tool of choice for tracking and filtering the motion of an object based on a series of position estimates made using the received radio signals. It is particularly useful when the dynamic behaviour of the object being tracked is known because then the Kalman filter can be optimised to filter the motion within the constraints of the way the object behaves.

Another increasingly important tool for doing motion estimation is the Particle Filter. Particle filters are gaining popularity as the cost of computing becomes cheaper. They require far more computational power than Kalman filters. Like Kalman filters particle filters are recursive so measurements can be processed as they are made.

A particle filter is a sequential Monte Carlo algorithm. It is used to track the positions of the object over time. A probability density function describing the probable motion of the object is constructed. As each observation (measurement) is made it is used to generate a number of probable locations that could result. These are the particles. The filter tracks many particles each representing probable locations updated recursively using successive measurements. The least likely locations are eliminated and new particles generated in their place. At any one time the most likely position (particle) is reported as the estimated position of the object. A particle filter is able to model non-linear systems and probabilities more accurately than the Kalman filter, but at a much higher computational cost.

Kalman and Particle filters are extensively covered in many good books such as Ristic [43] to which the reader is referred for further information.

8   When things go wrong

8.1 Systems, probability and false positives

8.1.1 Localisation and positioning are probabilistic

Any system for determining position or location is probabilistic in that the position it computes is based on noisy measurements and uncertainty. There is no such thing as absolute certainty or guarantee of the output. Therefore any application utilising the output of the system should take this probabilistic nature into account. All too often applications have not done this adequately leading to a level of scepticism amongst users, and in some cases mistrust of the systems.

8.1.2 Locating objects requires a system

In most 'professional' uses of locating systems it is necessary to place a level of trust in the system and to have confidence that it is correct. For these applications it is often useful to quantify the required system performance:

- What is meant by a location or position? Is it occupancy of a 'zone' such as presence in a room, a parking bay or a particular stretch of road? Is the position static or described by a sequence of positions (trajectory)?
- What is the duration of the location 'event'? What defines the start and end times of the event? How long do we have (realistically) to generate a location output?
- What is the acceptable level of false positive reports? A false positive is when the system reports that an object is in a particular location (for example room or parking bay) but it isn't.
- What is a tolerable level of failed location reports? This is when the system is unable (or unwilling) to output an object's location. This is generally a less severe error condition in which the user is told that the

system does not know where the object is and has to resort to backup or alternative methods for obtaining the position.

Having defined the system performance, appropriate technology selection and system design can be completed. When done properly this process usually leads to trustworthy and useful systems for positioning or locating objects. Unfortunately, with rapid proliferation of locating systems and technology recently, these finer points of understanding have often been missed. This has not been helped by manufacturers exaggerating performance claims (particularly accuracy) without quantifying the statistical and test basis of their claims. Furthermore many of these systems fail to provide a confidence metric describing the estimated accuracy and quality of the computed location.

There are some really important, and oft forgotten, questions applications developers should be asking when selecting a locating technology and system for their application:

1. What is the basis of the claimed accuracy? How was the figure arrived at? Under what test conditions? Can the vendor offer quantified performance figures under the conditions and configuration in which the system will be used for the target application? Is it possible to undertake a set of trials and tests to quantify actual performance in the real environment?
2. Does or can the system deliver a quality metric or confidence figure associated with position output that the applications developer can use to influence the way in which the data is used in the application? (Number of satellites in view and DOP are examples of quality metrics that are not particularly useful.)
3. What happens when the system fails to generate a position? Does it remain silent? Output the last known position? Indicate the failure and reason thereof?

8.1.3 False positives and failures

False positives represent the most serious failure mode of the positioning system. They arise when the system reports a position which is wrong.

For example a system might be designed to indicate when a particular object is in a particular zone. This may be a vehicle in a particular parking or loading area, or it may be an emergency worker in a specific room of a building, or it may be the position of an athlete in a race. Getting the position wrong and reporting the wrong position as fact represents a failure that could have serious consequences for the application.

On the other hand we have a failure to generate a position output. Compared with a false positive this is usually a less serious situation. For example: if instead of reporting that an object was located in a particular zone the system reported its position as unknown, the application could deal with this in a very different way. For example it may resort to a backup method of finding out the position, or it may postpone the action to be completed, or it may send a backup resource to support or verify the situation.

False positives therefore represent a failure condition that is not usually detected until it is too late and results in an error that may have costly consequences to the application. Since the performance of a positioning system is statistical and one cannot absolutely guarantee its performance there is a finite probability of false positives – as indeed for most things in life. It is up to the application developer and user to decide where to set this probability threshold. More accurate more reliable systems cost more, so the decision is usually a strategic one of balancing the cost of the system against the loss that is incurred through false positives.

On the other hand failure to generate a position is often a detected fault, and certainly represents a situation in which action specific to the determined position is unlikely to take place. They therefore usually represent a loss of opportunity or productivity rather than an error with more serious consequences. Generally a higher failure rate can be tolerated than the false positive rate, and most professional applications will offer means for dealing with the situation. For example the system is likely to know that a vehicle has arrived on site (either from the positioning system, or because the driver has reported the fact) but it may be unable to determine which loading bay it has parked in. In this case the system can generate an exception condition which triggers a manual search for the vehicle or a conversation with the driver.

Clearly different applications have different tolerance levels for false positive location reports and failures to output a position. They also have very different operating durations and response times and indeed may have other factors not mentioned here influencing the system design. The intent in this chapter is to raise awareness for applications developers about failure modes to prompt them to think in a more structured way about the failure modes and consequences of using positions from a locating system in their application.

8.2 Multipath

One of the most challenging problems to deal with in systems for location and positioning is multipath. In actual fact the problem is wider than just multipath and in this discussion we include any non-line-of-sight[1] propagation even if the non-line-of-sight path is the only received signal. Non-line-of-sight paths include refracted and diffracted paths.

In high precision systems propagation through materials other than air may need to be accounted for since the speed of propagation of the radio signal is different from air and therefore the distance travelled per unit of time differs.

In GNSS the signal from the satellites travels through the ionosphere and (significant) delays introduced by this layer of the atmosphere have been widely studied and techniques to reduce this effect are implemented in most modern GPS receivers.

Delays are, however, not the same as multipath and non-line-of-sight propagation which are arguably the biggest environmental factors that need to be dealt with in any terrestrial positioning system. They are also a significant cause of errors in GNSSs, particularly when GNSS receivers are used in cluttered urban and indoor environments.

There are a number of strategies for dealing with multipath, a few of the main ones are described briefly in the following sections.

---

[1] Line-of-sight does not, in this context, mean that there is an optical (visible) line connecting the transmitter and receiver; it is a path over which the radio signal can propagate representing the shortest distance between transmitter and receiver. This path may pass through objects that are transparent to the radio signal, but which may be optically opaque.

8.2.1 Looking for the earliest arriving signal

Since the direct path is the shortest, it is the first to arrive at the receiver, and therefore measuring the earliest arriving signal improves the likelihood that the measurement is of the direct path signal. Of course if there is no measurable line-of-sight signal it is not possible to measure it.

The channel response illustrated in Figure 4.7 shows an example of a received signal comprising several delayed versions of the transmitted signal. For the purpose of communications one aims to sum the various components using, for example, a rake receiver, to give the highest achievable S/N ratio in order to decode the data carried by the signal. For positioning the receiver tries to measure the time of arrival of the earliest arriving signal.

Whilst this technique works for broadband signals from which one is able to extract an estimate of the channel impulse response it is not possible in phase measuring systems. However, there are strategies that can be applied to phase measuring systems: for example the PLL (phase locked loop) used to track the decoded signal can be biased to track towards early phase measurements. This is relatively straightforward in systems in which the channel spread, or time difference between early and late arriving signals is small compared to the wavelength of the signal phase being tracked. In the event that the difference is larger than the wavelength the tracking problem is complicated by the need to track cycles in the time offset measuring function.

8.2.2 Building an overdetermined position solution

As described earlier a minimum of three measurements are usually required to solve for unknown x, y and time offset. When more than three measurements are available, the classical approach tries to construct multiple possible solutions using different combinations of received signals from which the likely position is selected.

However a far better approach is to use a minimisation technique in which a set of equations utilising all the available measurements is constructed and the best solution is found using numerical techniques based on minimising the function cost. Recapitulating the summary cost function from Section 4.5.3 for OTDOA systems, although the principles are applicable to others too, we have:

$$F_c(x, y, t_m) = \sum_{i=1}^{n} [R_i - r_i]^2. \tag{8.1}$$

The cost function is constructed as the sum of squares of the differences between the observed range to each of n neighbours and the actual range. In this example the solution is for the two-dimensional case, so we have three unknowns: x, y and t. The minimum number of measurements required is three, so four or more measurements would be used in an over determined solution.

Using more measurements than the minimum required leads to significantly improved solution robustness. The additional measurements used in the over determined set of equations lead to a total function cost that is broadly representative of the quality of the solution and which can be used, with care, to derive an estimate of the accuracy of the resulting position.

This approach results in statistically better results by using more measurements than the minimum required. The improvement in performance roughly follows a square root law which means that good improvement can be achieved with 10 to 20 measurements, but beyond this computational complexity (and system cost) rises disproportionately with the benefit achieved.

8.2.3 Received signal quality measures

It is also possible to use tell-tales from the signal measurements to infer how badly a signal is affected by multipath. The tell-tales might include:

- RSSI (received signal strength indication), or more particularly the time-varying nature of RSSI which is indicative of slow and fast channel fading as a result of multipath.
- Data decoder quality indicators such as error rates, symbol jitter and eye structure.

- Amount of noise in carrier phase tracking loop.
- Out-of-band and adjacent channel signal/noise levels.

These signal quality indicators can be used in different ways. In the extreme one could use the quality indicator for a simple go–nogo decision of whether to use the measurement in the solution set or not. In a more sophisticated implementation the quality measures could be used to determine a 'value' associated with each measurement in the measurement set.

8.2.4 Range residuals

Assuming that one solves the positioning equations based on more measurements than the minimum required using a suitable numerical minimisation method, this leads to a final function 'cost' which is the minimum value of the function being minimised. This function 'cost' is a useful indicator of the quality of the solution. However, one has to first normalise it. This is usually done by taking the square root of the cost scaled by the ratio between the total number of measurements and the excess number of measurements present:

$$f_c^{\text{norm}} = \sqrt{F_c \cdot \frac{n}{n-3}}. \tag{8.2}$$

However, when one considers the residuals it is possible in many cases to derive a lot more information about the specific solution. The residuals are the individual components of the vector $\mathbf{R}$ whose values were squared and summed to give the overall function cost:

$$\mathbf{R} = \left[(R_i - r_i)^2 \right], i = 1 \dots n$$
$$F_c(x, y, t_m) = \sum_{i=1}^{n} \mathbf{R}_i \tag{8.3}$$

The vector of residuals represents the individual contributions of each measurement to the total solution cost. The largest residuals therefore indicate which measurements are furthest from the final estimated solution.

Having computed and examined the residuals, there are various things that can be done. One might for example discard the measurement (or a few) with the biggest residual(s) and recompute the position using the remaining measurements. How well this technique works depends on how overdetermined the problem is and the nature of the errors. It is also dependent on the geometry of the measurements: skewed geometries can exaggerate some residuals at the expense of others. In all cases, care is needed, and as for all things, blind application of a recipe can lead to undesirable results.

8.3 Vulnerabilities and limitations of GNSS

This section outlines some of the special limitations and vulnerabilities of satellite navigation systems. It does not deal with conventional error sources such as ionospheric delays or errors in satellite clocks or navigation data – there are many excellent sources that deal with these factors in great detail [18] [1] amongst others to which the reader is referred for more information.

Other positioning and locating systems suffer from some or all of the problems described in this section too, although it is the widespread adoption of GPS that has led to these topics finally being given some of the attention they deserve. There is no doubt that many of these problems are not only real and serious, but they are likely to be a major factor in the uptake of location-based services and applications.

8.3.1 Loss of signal reception

As described in Chapter 3 the signals used for satellite navigation are very weak when they reach the surface of the Earth. Even with modern state-of-the-art high sensitivity receivers there is only about 15 dB of margin for signal acquisition and 30 dB of margin for signal tracking. These small margins mean that even with relatively little occlusion of the sky it is not possible to receive the satellite signals and therefore navigate using them.

There is a further potential problem with GPS signals which are based on CDMA (code division multiple access). This technique uses different codes for the different satellite signals and the receiver is able to separate them by correlating with the respective codes for each satellite. The amount of separation between different codes depends on the length and strength, specifically the autocorrelation and cross-correlation characteristics, of each code. The GPS C/A codes are relatively short codes, and although they have excellent auto and cross-correlation properties the code length means that a satellite signal weaker than about 20 dB below another cannot be reliably decoded and tracked. The implication of this is that even though a receiver may be capable of tracking signals down to −160 dBm, if one of the satellites is in direct view and received at, say, −129 dBm it will obscure signals below about −149 dBm.

There have been numerous measurements and tests done to establish the attenuation different materials have on radio signals, although most tests have been done for mobile cellular signals around 900 MHz and 1800 MHz and for ISM signals in the 2.4 GHz band. As a rule of thumb a basic brick wall is likely to attenuate the signal by 6 to 9 dB and a reinforced concrete wall (or floor) is likely to lead to around 30 dB of attenuation. Therefore GNSS signals are unlikely to be receivable in an indoor situation in which one or more floors are above the receiver position − unless the satellite signals are receivable through windows and walls.

Relying on signals received through windows and walls will inevitably lead to a worse navigation solution, for several reasons:

1. The DOP is likely to be poor due to a skewed visible satellite constellation. Since the signals are coming through the side of the building, and generally from only one or two directions, the satellites being received are likely to be lower in the sky and predominantly off to one side.
2. The received signals may suffer from multipath. In many cases the satellite signal being received is via a bounce path off an adjacent building. This has the effect of skewing the navigation solution. Although the satellite signals are circular polarised which helps to

reduce single bounce reflected signals, multipath signals from satellites are still a significant issue in indoor location.

3. The signal from a satellite directly in view, perhaps through a window or other opening in the building, may obscure some of the weak satellite signals being tracked.

Once the receiver has lost reception of the satellite signals, it needs to go through a reacquisition phase before resuming navigation. If one or more signals continue to be tracked even though there are insufficient for navigation, the receiver can make assumptions about movement during this period which, if valid, can lead to very quick signal reacquisition.

8.3.2 Interference

With increasing number of radio devices and use of wireless technologies there is considerable pressure on radio spectrum. It is inevitable that radio devices will sometimes interfere with one another. Interference may be classified into two kinds:

- In-band interference. This is interference that lies in the same band as the signals of interest, e.g. GPS L1 or L2 bands.
- Out-of-band interference. This is when signals outside of the band of interest affect the performance of the radio receiver.

Since GPS signals are so weak it is also susceptible to interference caused by solar activity. Solar flares can lead to geomagnetic storms which some predict could affect GPS operation. Solar activity follows an eleven-year cycle approximately and 2013–2014 will be the peak (highest) level of activity in the current cycle. So far no significant effects on GPS have been observed.

8.3.2.1 In-band interference

Interference signals are usually unintentional; the situation when they are intentional is a special case and is dealt with in detail in Section 8.3.3.

No radio transmitter is perfect and therefore in addition to the wanted signals it is transmitting, such as cellular mobile signals around 900 MHz or 1800 MHz, it also radiates signals outside of its operating band. The level at which it is permitted to emit these signals is governed by the radio regulations for its operating frequency. For example a 3GPP base station transmitter has a mask level of −30 dBm for the 1 GHz to 2 GHz (approximately) band. Whereas signals are generally well below this level, in the event that a base station was close to the allowable mask in the GPS band around 1.575 GHz it will have the effect of blocking GPS reception when the GPS receiver comes close to the base station.

Another potential source of unintentional interferers is harmonics of digital clocks in electronic equipment such as computers. Clocks operating close to 1.57542 GHz or for which this is an integer multiple of the clock frequency can and do generate signals that may interfere with GPS receivers. This is a particularly challenging problem for equipment that contains both GPS receiver and other radio or computing circuits, such as mobile phones, because they are inherently in close proximity to one another.

In-band interference in the 2.4 GHz ISM band is particularly problematic given the intensive use of this spectrum by many different competing systems using many different protocols including: IEEE 802.11 (Wi-Fi), IEEE 802.15.4, ISO 24730 (RFID), Bluetooth and others. Particularly problematic for many users are the analogue wireless video devices that operate in this band. Microwave ovens operate at 2.5 GHz and can cause interference too.

8.3.2.2 Out-of-band interference

Out-of-band interference arises when a radio signal outside the band of interest causes the GNSS receiver to malfunction. This may arise from simple overloading of the radio receiver, or because of non-linearities in the receiver or causing the out-of-band radio signals to mix destructively with internal clocks and oscillator frequencies.

Out-of-band interference as an issue for GPS arose recently with the LightSquared situation in the USA.

According to the company: '*LightSquared will unleash the boundless opportunity of wireless broadband connectivity for all. We believe that it is time to transform the broadband industry to one that truly fosters innovation, creativity, and freedom of choice – with limitless and un-imaginable possibilities.*'

LightSquared controls a block of the United States spectrum (1525–1559 MHz) in the L-Band. It received FCC authorisation in November 2004 to use this spectrum to build a nationwide wireless broadband network. In January 2011, LightSquared received a conditional waiver allowing terrestrial-only use of this spectrum, which would result in a large number of base stations being deployed. The spectrum allocated for the GPS L1 signal is 1559 to 1610 MHz. This resulted in significant backlash from many GPS users who claimed that the LightSquared signals would interfere with GPS receivers. This interference could arise from in-band interference arising from the LightSquared base stations, but, perhaps more critically, tests showed that many receivers on the market would fail in the presence of strong out-of-band signals so close to the receivers' operating frequency.

Proponents of LightSquared argued that it was the responsibility of receiver manufacturers to ensure that their receivers were resilient to out-of-band signals, but GPS users argued that there were numerous receivers in the market place which had been designed for the environment into which they were sold, and furthermore that adding the required extra filtering would increase the cost of new receivers.

Despite many proposed changes from LightSquared, the FCC ruled, after conducting a number of tests, that LightSquared was not permitted to deploy its proposed terrestrial system. The FCC felt that the risk of interference to general GPS users, the aviation industry and military users was too high. At the time of writing LightSquared has filed for bankruptcy to give it time to restructure.

Out-of-band interference is also a problem for ground based wireless systems operating in the 2.4 GHz ISM band. 3G bands are located fairly close to the ISM spectrum: 2.1 GHz is already widely used and spectrum in the band 2.6–2.7 GHz is presently being licensed in many countries. Both can affect the operation of some 2.4 GHz products.

8.3.3 Jamming and personal privacy devices

8.3.3.1 Background

The increasingly widespread use of PPDs (privacy protection devices) in recent times has been the topic of a number of recent conferences, workshops and press articles. A PPD is a small radio transmitter device designed to transmit a radio signal in the GNSS band, and often in other critical communication bands for mobile cellular services, the objective being to jam nearby GNSS receivers to prevent them from being able to determine a position. These devices are typically used by fleet vehicle drivers and private drivers in company cars when they don't want their employer to be able to track their whereabouts.

Using a PPD to jam GPS signals is illegal in many countries including USA and throughout Europe, but it is not illegal to possess one. They are readily available and can be purchased from a number of international suppliers via the Internet. They are small, inexpensive (< $50) and typically plug into the auxiliary power socket in the vehicle. A google search for 'gps jammer for sale' under images turned up 338 thousand hits.

Unfortunately these devices can have consequences far beyond their intended use. Even a low-power jammer can block GPS reception over large areas – ranges of up to tens of kilometres in some cases. Not only do they affect vehicle navigation systems and other personal GPS receivers, but they can affect aircraft landing systems, mobile cellular networks and military operations.

In January 2007 the US Navy accidentally left a GPS jammer operating on a ship docked in San Diego harbour. *New Scientist* magazine reported on this incident in its March 2011 edition as follows: '*It was just after midday in San Diego, California, when the disruption started. In the tower at the airport, air-traffic controllers peered at their monitors only to find that their system for tracking incoming planes was malfunctioning. At the Naval Medical Center, emergency pagers used for summoning doctors stopped working. Chaos threatened in the busy harbour, too, after the traffic-management system used for guiding boats failed. On the streets, people reaching for their cellphones found they had no signal and bank customers trying to withdraw cash from local ATMs were refused.*' A number of people contend that the report exaggerated the consequences of the incident which was not as serious as made out. Whatever the truth of the story, GPS signals can be easily jammed and the spiralling use of GPS jammers and PPDs is a cause for serious concern.

Of particular concern is the potential for their misuse, rather than the more innocent desire to simply prevent one's employer from tracking one.

8.3.3.2 The official response

A consortium of navigation experts in the UK has recently completed a TSB (Technology Strategy Board) funded project called GAARDIAN (GNSS Availability Accuracy Reliability and Integrity Assessment for timing and Navigation) which developed and tested a sensor which can *'continuously log substantial amounts of data about the GNSS signals of interest'*.

This is now being followed by a project called SENTINEL (SErvices Needing Trust In Navigation Electronics Location and timing) which aims to enhance the sensors so that sources of interference can be pinpointed and which will help officials to determine whether the source is accidental, natural or a deliberate attempt to jam GNSS signals. Whilst the main aim is to protect critical national infrastructure, the work has highlighted some interesting trends:

- the use of PPDs is on the rise, although they are not yet widespread in the UK and certainly nowhere nearly as ubiquitous as, for example, the Guardian Angel products used to identify and warn drivers of speed camera locations (also illegal to use);
- PPDs can cause significant impact on GPS receivers and GPS jamming is a real issue;
- many of the project findings are classified and may not be published publicly.

In the USA the Department of Homeland Security (DHS) has launched a programme to study and assess the risks to GPS services. The intention is to develop a plan to mitigate interference. The plan is called the National Risk Estimate (NRE) and, like the UK work, is concentrating on the risk to critical national infrastructure. The NRE will weigh up whether and how disruption of GPS signals would affect five key sectors: banking and finance, communications, emergency services, energy and transportation. This work is following four strands:

1. assessment 'to identify current and possible future ways to mitigate' against disruption of GPS signals;
2. development of an incident reporting portal;
3. development of a sensor system called Patriot Watch that can be used to monitor for and detect incidents;
4. prevention of hostile use of GPS in the USA.

Even though it is presently illegal to transmit on GNSS frequencies, the mood of the authorities tends to be one of enacting stricter and more far reaching laws to punish those who do:

- In 2009 regular outages at Newark Liberty International Airport were finally traced to a 200 mW PPD used by a truck driver as he passed the airport each day [44].
- In July 2010 two truck thieves were jailed for 16 years in the UK – they used GPS jammers to prevent the trucks from being tracked after they stole them.

Whilst this approach may deter some users, the reality is that it is extremely difficult to locate a PPD in order to prosecute the user. If anything the tough line being adopted by authorities is making it harder for researchers to work in the area, and could potentially set back research efforts to develop real mitigation techniques in receivers.

8.3.3.3 Impact of a jammer on GPS signals

A number of researchers have produced models and estimates on the effective range of a GPS jammer. *GPS World* [45] reports on a series of tests which were combined with models in which the researchers looked at and measured the signals emitted by 18 different jammers on the market. They concluded that the weakest of the jammers had a range of 300 m for tracking and 600 m for acquisition, and the strongest jammer affected tracking at up to 6.5 km and acquisition at up to 8.5 km. The tests were conducted using standard COTS (commercial off the shelf) GPS receivers.

InsideGNSS [44] presents a well-argued case for the risks of GNSS jammers on aviation and uses the Newark Liberty International Airport incident as a reference case. The article includes details of selected jammer signals and looks forward to steps and approaches that might be used to protect aviation services from accidental or intentional jamming.

North Korea has acquired powerful GPS jammers which it uses to jam GPS signals in neighbouring South Korea. It has been reported that they have a jamming range up to 100 km.

The Lighthouse Authorities of the United Kingdom and Ireland [46] carried out a GPS jamming test off the coast at Flamborough Head in 2008. During the tests they operated GPS, DGPS (differential GPS) and eLoran receivers. Both GPS and DGPS receivers were affected by the jamming position. Rather disturbingly they did not simply fail to generate positions they reported positions that were incorrect by up to 20 km without any indication of the error. eLoran continued to operate correctly throughout the jamming exercise. The impact of the GPS failure would have had serious consequences for shipping in the area were it not known to be a test situation.

8.3.3.4 *Techniques for mitigating jamming in the receiver*

Most of the commercially available GPS jamming devices operate in a very simplistic manner: They generate a swept signal, generally with a sawtooth time structure spanning the band they are jamming. However, unintentional interference signals are often narrowband tones arising from oscillator harmonics present in the device.

There are a number of techniques that can be used to detect and/or mitigate the effect of the jamming signal:

- Antenna pattern – particularly for fixed installations as used in airports and for mobile cellular base stations and to some extent for receivers permanently installed in vehicles, an antenna pattern providing good attenuation of very low elevation signals can be an effective method of combatting ground-based jamming signals.
- Receiver front-end filtering can help eliminate unwanted signals. Using a filter only as narrow as the desired receive band helps prevent unwanted near-band signals from getting into the radio receiver.

A notch filter can be effective for eliminating tonal interferers such as oscillator harmonics.

- Simply looking at received power can provide an indication that a jammer is present, because jammer power is often far greater than the maximum power received from the satellites. It does not help mitigate the jamming signal, but as illustrated in the Lighthouse Authorities tests [46] the GPS receivers used in the test did not detect malfunction.

Considerably more work is needed to research techniques for jammer detection and mitigation. Due to the onerous legal implications of operating devices that transmit jamming signals, and restrictions about publishing findings, working in this area is a real challenge for researchers.

8.3.4 Signal spoofing

Whilst jamming of GPS signals as described in the previous section is a problem, the real threat for hostile attacks against GPS is signal spoofing.

Consider the implications of being able to generate a set of false GPS signals that when received by a GPS receiver cause it to output a position indicating that it is somewhere else. This would allow the spoofer to manipulate nearby GPS receivers to appear to be at a position desired by the spoofer.

Nottingham Scientific with the SkyClone system has demonstrated the feasibility of generating a set of GNSS signals matching any desired time, place and situation, including signal obstructions and presence of jammers. These signals are generated to appear authentic to present-day GPS receivers.

In order to mitigate the potential threat of signal spoofing it is essential that the receiver is able to authenticate receiver satellite signals. The demand for being able to do this has been rapidly escalating in recent years, mainly to protect the critical national infrastructure uses of GPS.

A number of proposals following two main approaches have been put forward for signal authentication:

- user segment authentication leveraging existing services in order to detect spoof signals;
- signal authentication services integrated in the GNSS itself.

The first can work with existing GNSSs that don't offer authentication services but the second requires new systems and/or modifications to existing systems and architectures. An example of a proposal fitting the second category can be found in *InsideGNSS* [47].

Whilst spoofing GNSS signals is feasible it is rather complicated to do properly and is still very much the domain of a limited number of industry experts. However, it is only a question of time before those with hostile intent will be able to deploy devices capable of spoofing GNSS signals.

8.3.5 Lack of quality metrics for the fix

The industry standard interface for GPS receivers to output location related data is specified in the NMEA 0183 standard [48], although a number of other protocols also exist. NMEA 0183 makes provision for a status field called fix quality in the GPGGA sentence (Global Positioning System Fix Data). This field defines whether a fix was obtained and the method used to obtain the fix. However, to call it a fix quality indicator is stretching the point somewhat.

For applications developers the lack of real fix quality information from GPS receivers is a serious limitation. What is needed is a measure of how trustworthy the position fix is, and an estimate of realistic expected accuracy would also be valuable to many applications. Very few GPS receivers make any attempt to estimate true quality metrics, and even fewer report the metrics to applications that use the location information from the receiver.

There are, however, a number of factors that applications can take into account to formulate an estimate of the quality of the position fix:

- Number of satellites in view – fewer satellites mean that there is less redundancy in the navigation solution, but on its own this does not say much about the quality of the fix.
- DOP – provides a single parameter representing the way in which the satellites are distributed around the solution, and acts as a multiplier to the expected accuracy.

- Variation in satellite signal strengths – if all satellites have similar signal strengths the probability is that there is a fairly homogenous view of the sky, whereas large variations between satellite signal strengths may suggest obscuration and potentially multipath degradation.
- Signal-to-noise ratio for satellite signals – if available can help assess quality of the measurements.

It is up to the application developer to decide how to use this (partial) information in order to assess probable navigation accuracy. It would be extremely useful for the GPS receiver to output a simple quality metric providing a best estimate of the accuracy of the fix. Since the receiver has access to so much more information about the quality of the signal it is possible to compute an estimated performance most of the time, provided that more than the minimum required number of satellite signals are available.

The same principles go for most other radiolocation systems. Very few provide quality information for use by the application developer. However, one or two systems do present useful quality metrics. Sometimes these take the simple form of: *excellent, good, poor, none*. Some systems are able to provide an estimate of the accuracy of the fix along with a confidence in the fix. Some are even able to provide full error ellipsoid parameters – see Section 7.1.4.

8.4 Strategies for dealing with failure

Positioning failures will happen; by its very nature the computation of position is statistical, being highly dependent on many different environmental factors and measurements of noisy radio (usually) signals.

There are three stages to dealing with positioning information:

1. Detection of failures or faults;
2. Failure mitigation;
3. Use of backup methods or operating procedures – i.e. dealing with failures.

In Section 8.1.3 the failures were classified as either 'false positives' or 'failures'. Whilst both are failure modes the distinction is that those in the first category are reported in good faith as being correct and are acted on accordingly, whereas those in the second category result from either no location being reported or the reported location being tagged as untrustworthy or potentially untrustworthy by the system.

Depending on the cost to the user application the design of the positioning system needs to ensure that a sufficiently low probability of undetected failure occurs so that the system benefit outweighs the cost of failure. In general we should strive to minimise the probability of undetected failures occurring. This doesn't necessarily mean that the system fails less often, just that we are able to detect when a failure occurs. A crucial input to this process is the availability of proper quality metrics from the location system as discussed in the previous section.

When the system generates a position that is thought or known to be untrustworthy, it may be possible to seek ways to mitigate the fault or uncertainty. This could be by gathering additional position estimates from the system or by delaying the location report while further analysis is undertaken.

The final and crucial step is to have in place fallback or backup methods for obtaining location information. In the Lighthouse Authorities study into GPS jamming [46] they recommended that eLoran should be used as a backup system to GPS for marine navigation. In this situation the strength of eLoran is that it has different failure modes to GPS and is not likely to be affected by the same interferer as GPS.

Alternative backup methods might be manual methods for obtaining position, or changes to the operating process appropriate to the situation. For example one might have to revert to reading a paper map. In E-911 cellular systems fallback from GPS or UTDOA to Cell ID is used.

9 Location-based services and applications

9.1 Essential principles underpinning services and applications

9.1.1 Navigation

There are relatively few applications for which location or position is the core purpose. Navigation is the notable example in which the purpose is to find a place or navigate a route to a place. Before the advent of modern technology ocean navigation was a huge problem facing mariners at sea. With the help of a sextant measuring the inclination of the Sun above the horizon at midday allows one's latitude to be estimated. However, determining longitude is far more difficult. Sobel [9] describes the impact of this problem and efforts to solve it in the days of the early great sea navigators.

Given modern GPS systems it is easy to determine one's position to good accuracy outdoors when there is an adequate view of the sky. However, the application challenge is to provide guidance to the user leading them along a route from one point to another.

One hears many stories and jokes about GPS navigation systems leading users astray. Whilst this is occasionally for technical reasons – failure to obtain a fix, or positioning errors – more often than not the problem lies with the navigation aspects: for example old or incorrect map data or lack of clarity in the guidance instructions.

Satellite navigation is largely a solved problem, but there are still challenges for application developers to solve:

- Clearer guidance instructions given in a way that more closely mimics the way users relate to their environment. For example it is not always easy to identify a road by its name, we are more likely to relate to an instruction like: 'Turn left just after the stone church with the tall spire', or 'go past the White Horse pub'.

- Improving the responsiveness of the system to changes in circumstances: traffic congestion, road works, planned closures, different weather conditions etc.
- Non-vehicle use, particularly for navigation on foot in urban areas.
- Indoor navigation and guidance.

9.1.2 Navigation on foot

Navigation and route guidance systems for use in cars simply don't work very well when on foot.

Whereas cars travel on roads which are comprehensively mapped and signposted and map matching algorithms can be used to fit the movement to roads, walking routes often take one across or through unmapped areas and are far less constrained.

Cars have wheels which constrain their motion in the direction the wheels are pointing. People can move in any direction and may be carrying the navigation device in a completely unconstrained way. Its orientation relative to the route may vary unpredictably.

Walking routes often take the person and navigation device into areas where conventional technologies such as GPS don't work – for example indoors.

Instructing a person to walk a specific distance and direction is not particularly constructive on its own; it is essential to direct them to a particular place that is recognisable and unambiguous with the distance and direction being the context.

Direction is one of the challenges for foot navigation. Whilst technologies exist for measuring the orientation of the sensor device, it is not possible to rely on a particular association between the sensor and the user. The relationship depends on where and how it is worn or carried, and this may change over time.

The other thing to consider is how these instructions are relayed to the user: verbally, visually, aurally, or even using some other haptic indication.

9.1.3 Indoor navigation

Navigation indoors and in built-up areas is particularly problematic. Firstly there is the technical challenge of obtaining a meaningful position; secondly describing the context of the position and giving directions is a challenge.

Conventional technologies like GPS simply do not work adequately indoors. If a position fix can be obtained it is usually too poor to be of use for reliable indoor navigation. And even if an adequate fix could be obtained presenting coordinates in latitude and longitude is not very useful because few indoor spaces are mapped in this way. Furthermore many buildings are three-dimensional having multiple floors, so excellent height resolution is essential – something that GPS is not particularly good at.

When maps or plans of indoor spaces are available they are usually in the form of a building plan, drawn in Cartesian coordinates, floor-by-floor, with an arbitrary origin and orientation, which is very seldom linked to a global coordinate system or even a local geographic mapping system, such as OSGB. Therefore for navigation around a building it is necessary to determine the position in the context of the building's local coordinate frame. Furthermore, the building plans will rarely indicate additional features like furniture and fittings, which may also move around from time-to-time.

So assuming that one can determine a position within a building the challenge is how to relay directions or guidance instructions to the user. Certainly these need to be presented as a descriptive context of the environment using clearly visible and identifiable landmarks.

The other challenge is that navigation to a fixed point is not necessarily what is required. For example one may wish to find and meet a friend or colleague who may also be moving around the area. One approach in this case is to guide both people relative to the fixed environment, but, especially when getting nearer, it may be preferable to provide guidance as a simple vector – direction and distance – obtained by making a relative positional measurement between the two.

9.1.4 Cooperative networks

There are many situations in which cooperative or relative positioning is the essential aspect of the application rather than necessarily describing positions in a fixed global coordinate frame:

In a navigation context being aware of the relative flow of traffic on a road network can be used to give advance warning of changing road conditions or an emergency situation. Collision detection systems monitor the relative positions and approach speeds between vehicles, or vehicles and objects and use this information to alert the driver in advance to changing conditions ahead. This information may be best obtained by making relative measurements between the vehicles directly rather than referring both to a global coordinate frame and differencing the results – after all there is little benefit from knowing the absolute position.

Indoor navigation and navigation on foot is more often than not performed in a local context, such as within a building. For example the question 'take me to the toilet' is most concerned by how far and in which direction to walk, rather than knowing the exact locality of the toilets.

An application which links a shopper to the till at which they are making a purchase, could compute both positions and difference them, but in reality a far better approach is to measure the relative proximity of the two directly.

Many other applications benefit most from knowing the relationship between objects, so the application developer should consider whether it may be more appropriate to directly measure the relative positions (and orientation and velocity) of the objects than to independently compute their positions and compute the difference between these measurements in order to define the relationship.

9.1.5 Geofencing

A geofence is a virtual perimeter for a physical geographic area. It delimits a region in space in three dimensions defining the boundaries of this region. Examples are: a room, or zone within a room; a neighbour-hood; perimeter of a property or business; or simply a radius around a particular point. In this book we view a geofence in this broad interpretation. Unfortunately it does sometimes have a negative connotation and has been used to describe a virtual boundary imposed on workers by an employer, the intention being to prevent them from leaving an area, or to penalise them for doing so. In our view this is a very restricted interpretation of an essential and fundamental principle used in many applications that use location information.

Geofencing forms an essential part of almost all the location-based services and applications described in this chapter; after all the zone bounded by the geofence is the essential context used by the application – most applications are based on a logical context for the location and using a geofence is a way of contextualising the position and making it useful to the application.

Being able to configure a geofence that can generate an alert when a user enters or leaves an area and thereby trigger an application associated with the area is a key hook into useful location-based services and applications. Equally being able to determine whether a user is at a particular location (defined by the geofence) is fundamental to many applications.

9.2 Location-based services

Location-based services (LBS) is a widely used industry term referring to a whole raft of different applications, largely in consumer space, and largely exploiting common consumer devices such as mobile phones, in which a service or application is either enabled or enhanced through the addition of location information. The range of these potential services is almost as great as imagination allows.

The essential ingredients facilitating the growth in consumer LBS are:

- availability of devices containing free locating capability; i.e. in which the user does not have to pay a fee to a service provider in order to obtain their position. GNSS is the main technology used for locating devices, but cellular positioning and techniques such as the use of Wi-Fi access points pioneered by Skyhook, are also frequently used.

- ubiquitous and free or low-cost access to maps and geographic information. Google arguably has a monopoly on maps and GIS data but other service providers like Bing (Multimap) and open source alternatives like openstreetmap.org are also available.
- mobile data services allowing users to access online databases and sources of information.

The following is a quick non-exhaustive look at some of the emerging applications and services that could drive the growth of consumer LBS.

9.2.1 Google Latitude

Google maps and location services are made available to mobile users with smart phones, including Android, BlackBerry, iPhone, iPad, Symbian and Windows Mobile devices. They are able to make use of the internal GPS receiver, cellular base-station information and even Wi-Fi access point locations to determine the location of the device. This information is linked back to maps and the locations of other users for a range of applications including: turn-by-turn navigation on the road or on foot; whereabouts of friends and colleagues; location of nearby services etc.

9.2.2 Location sharing, friend finder and similar

One of the main aspects of Google Latitude is the ability to share one's location with friends and colleagues, and to see where they are. This is a major growing area and there are many other applications that offer this capability to smart phone users:

- 'Find My Friends' is an Apple application for iPhone and iPad users, intended to rival Google Latitude.
- There are numerous independent apps for smart phones designed to mimic Google Latitude, each with a slightly different set of additional features: Footprints; foursquare; Gowalla; Glympse; Viewbubble; ItSpot; and many others.

9.2.3 Nearest service

This is another major category of applications on smart phones and mobile devices. The essential proposition is to be able to search for a service or facility nearby. These range from looking for restaurants, theatre, filling stations, supermarkets and indeed almost anything one can think of. The range of available apps in this category is extensive.

9.2.4 Traffic services

Real-time traffic updates for car navigation systems are available but they are still less than optimum, and further improvements are likely over the coming years.

9.2.5 Travel and leisure guides

Electronic travel and leisure guides that are sensitive to one's location and which automatically provide the appropriate commentary are becoming available. City-wide guides and guides for large open-air areas often use GPS as the locating technology. Indoor guides, such as museum and gallery guides are generally based on detecting presence within a room. There is clear scope for enhancing these with full positional knowledge of the customer – including their exact position and orientation (for example which painting they are standing in front of).

9.2.6 Push advertising

Many experts in the field believe that push advertising based on the location of the recipient is the way things will go. After all services need to be paid for somehow and by someone. Google and many other Internet services are funded from advertising and sponsorship revenue. If advertising can be targeted more effectively based on the location of the recipient it is likely to be more appropriate and therefore more effective at generating revenue and more palatable to the user.

Just like road pricing, push advertising polarises opinion in very fractious ways. Proponents argue that having targeted advertising is good for advertisers and the recipient since it is relevant, arguing that users want to know that, for example, the nearby coffee shop is running a special deal at the moment. Opponents, on the other hand, argue that it is just more commercial chaff to annoy us and often cite issues around privacy as a concern; after all, does the coffee shop have any right to the recipient's location data?

9.2.7 Sport and fitness monitoring

Another area in which numerous applications have emerged is in fitness and sport monitoring, mainly for walking, running and cycling but also other outdoor sports. A range of different apps which count steps, track the route followed, measure speed and estimate energy use have emerged for use on popular smart mobile phone platforms which have GPS and motion sensors, such as accelerometer and magnetometer fitted.

9.2.8 Location-based games

Location-based games are games with a location context played in a real setting, combining real and virtual worlds. Some games developers believe that location-based games will be the next killer app for smart phones and wireless-connected devices with locating and positioning capability. Some of these games are already beginning to emerge, for example Shadow Cities. Angry Birds will also be location enabled soon.

It is taking a while for this area to mature, and for some time there has been talk about a virtual representation of the world being used to depict the real world in sport – such as for football, horse racing, sailing and motor racing, including the ability to have one's virtual character enter the virtual world and compete against the live participants. At this stage mixed and augmented realities are still somewhat embryonic, but with good location context (including orientation) technologies for providing a location feed into these applications are maturing.

9.3 Professional applications

9.3.1 Road usage pricing

Tolls have been levied on travellers since eternity, being described in Greek mythology in which the ferryman Charon charged a toll to carry the dead across the rivers Acheron and Styx to Hades.

Modern toll roads are now widely used in many countries of the world. They were introduced and continue to be built as a way of financing the building and maintenance of roads, bridges, tunnels etc. The whole subject of how to finance the construction and maintenance of roads is one vexing many governments of the day. Traditionally they have been built using taxes, but increasingly there is pressure to compartmentalise revenue and costs.

Several different approaches to collection of payment for roads and transport infrastructure are used, or are being proposed:

- direct levy on each vehicle through a licence or tax to operate the vehicle on the public roads;
- tolls in which payments are taken at specific points in the road, often to cover the cost of a particular section of road such as a bridge, tunnel or stretch of motorway, or covering a specific area such as the central London congestion charge;
- a levy or tax on fuel;
- a fee or tax based on time and/or distance spent on the road.

The latter has been the subject of extensive trials at different times and in different parts of the world [49] [50]. In principle it is the fairest method because users of the roads pay in proportion to the amount of use they make of them. It is also the one most dependent on being able to acquire accurate location information during vehicle use. Road pricing divides opinion like few other subjects. It is probable that at some time in the future we will see true road pricing implemented, but there remain many challenges to be solved:

- Is the location data reliable enough to use? Trials have shown that GPS in its current form is not sufficiently accurate for urban use, but simulations show that when taken together with Glonass, Galileo and/ or Compass systems it will be sufficiently accurate.

- Robustness of the information and collection method – whether in real time using a wireless communications technology, or some other method. What surety is there that the collected data has not been tampered with?
- Privacy concerns present a major obstacle – many road users do not wish their whereabouts to be tracked and recorded.
- Secondary information that can be obtained from the collected location information – perhaps simple speed monitoring, but equally the information may enable driver behaviour to be measured with, say, detection of driving under the influence of alcohol or other substances. Could this secondary information be used by law enforcement authorities, and if so what laws govern its use?
- What if a driver's road usage information fell into the hands of third parties outside of the official authorities tasked with collecting taxes?
- Robustness against jamming, spoofing and fraud – these are major problems when using solutions based on GNSS alone. However, when combined with enforcement options using DSRC (Dedicated Short Range Communications) and ANPR (Automatic Number Plate Recognition) systems, it is argued that the system as a whole will be sufficiently reliable.
- Social and cultural issues should not be underestimated. The motoring lobby is very powerful and cars carry enormous emotional power. The social and cultural issues could hold governments back from implementing road usage charging schemes for many years.

9.3.2 Healthcare

There are many potential uses of location and positioning information in healthcare: including care for people with dementia, post operative recovery and even to monitor and track drug usage after patients have been discharged from hospital.

Dementia is a huge problem facing our greying population. It generally afflicts elderly people and is a progressive disease for which there is no cure yet. It starts out with minor lapses of memory and forgetfulness and progressively gets worse until in the final stages the person with dementia cannot remember what they did 5 minutes ago and may not even be able to recognise their own family members.

It is possible to delay the onset of dementia, and research has shown that people living at home with their families deteriorate more slowly than those placed in institutional care. Throughout the entire progression of the disease there is scope to use technology to assist both carers and the person with dementia. Technology can monitor their whereabouts and raise an alarm should they leave the house without family members or carer knowing, or it can monitor for and detect falls. Clinicians can use behavioural information to chart the rate of decline of the person with dementia as the disease progresses and plan ahead for appropriate treatments that can slow the onset of late-stage dementia, thereby giving the person and their family a better quality of life for longer.

For people in post-operative recovery or undergoing physiotherapy it may be helpful to be able to measure mobility and locomotion by tracking position, distances walked, step count, step rates, etc.

There are also many applications of positioning and tracking technology within hospitals and care homes. These range from tagging and tracking equipment, beds and wheel chairs, to monitoring laundry; and of course knowing the whereabouts of staff (porters, doctors and nurses) may also be of benefit.

9.3.3 Time and usage based vehicle insurance

Known by many different terms including: Pay As You Drive, Usage-Based insurance, Telematics insurance, Black Box insurance, GPS insurance, Smartbox insurance and others, this is a service in which the cost of insuring a vehicle and driver is linked to when and where they went. It was pioneered in the UK by Norwich Union Insurance in 2006, although they no longer offer Pay As You Drive insurance policies. However, there are many companies offering this kind of insurance in many countries around the world.

The basic premise is that the more time a driver spends driving the higher the probability of an accident, especially when the locality and time of travel are taken into account. The service usually relies on tracking the vehicle position and speed against time using a GPS receiver, and reporting usage data to the insurer on a regular, typically daily, basis. In addition the 'black box' usually also records a number of behavioural parameters that can help to characterise the driver behaviour and the nature of an accident.

Whilst these systems can be advantageous to young and at-risk drivers and those with variable driving patterns, opponents question aspects of privacy. For example, suppose that the driver behaviour information was able to indicate that the vehicle driver was unwell, under the influence, or perhaps a different person was driving the car – would the insurer be compelled to reveal this information to the law enforcement authorities?

9.3.4 Child safety

The fear of what might happen to your child when they are out-of-sight is very real for most parents, and leaders of school and youth groups under their care, especially during trips away from home.

Many companies offering solutions to this problem have sprung up. Most produce a small device worn by the child – usually in the form of a wrist watch – which uses GPS to locate the child and a mobile data connection (such as GSM/GPRS) to report positions to the parent or carer. Some use local wireless technology, often based on IEEE 802.15.4 (a), to provide accurate localisation or a simple range representing a virtual tether.

The major drawbacks to GPS are:

- It works poorly, or not at all indoors.
- It requires an additional data connection such as GPRS.
- It is very power hungry.[1]

---

[1] Compare with conventional GPS receivers such as those made by Garmin or Magellan, which provide typically 10 to 30 hours of operation from two AA batteries (far too big for a wrist watch sized device).

- It is not sufficiently accurate: whereas one might get 10 metres outdoors, in urban and occluded areas with a body-worn device, >50 metres is more common, and worse when infrequent position updates are done to save battery life.

On the other hand devices based on local wireless networks only operate over defined areas, or when they are within range of the parent or carer. For younger children this is probably advantageous, whereas for older children one might choose to track the whereabouts of their mobile phone using its inbuilt positioning capability.

9.3.5 Worker health and safety

There are many commercial systems on the market for tracking lone workers for the purpose of protecting and managing staff when they're out of the office. Lone workers may find themselves in potentially hazardous situations, whether at risk from the natural environment or strangers. Typical applications range from health visitors to construction, oil & gas workers and underground miners.

They are usually provided with a tracking device that is able to locate them and which often includes a panic button which can be used to call for help. Some devices are also able to do automatic detection of 'man down'.

Different solutions use different technologies depending on the worker requirements:

- mobile phones may be tracked using the built-in GPS and mobile network connection, often in conjunction with an applet running on the phone;
- RFID identity badges may be used to record worker location when they report to different premises;
- private mobile radios ('black' radios) as used for professional services may be used in conjunction with secondary localisation techniques using Wi-Fi, Bluetooth, GPS or others;
- small bespoke devices using radiolocation and/or GPS.

One of the biggest challenges in workforce tracking is ensuring that workers do not forget to carry an operational device with them at all times. Key to being successful is solving the human factors in such a way that the solution is part of the worker's job and not an extra 'administrative' task they must add to their everyday duties.

9.3.6 Recovery of lost assets

Everyone must have at some time wished for something that would help them find the car keys or glasses that they have mislaid, or help them find where they've parked the car. These are some of the frivolous mass market opportunities for finding lost assets, but there are also many serious applications, particularly associated with emergency situations and rescue operations:

- locating missing people after a natural disaster: avalanche, earthquake, etc.;
- tracking down missing people after an accident, such as the sinking of a ferry or ocean liner at sea.

For some of these applications the missing person will not be carrying a tracking device, but even in these cases technology can help by accurately tracking and recording the rescue and emergency workers, thereby providing an accurate auditable record of the entire search and rescue operation.

9.3.7 Operations management

Whilst health and safety are often headline grabbers, it is the use of location and positioning information to improve productivity and operational efficiency that is really driving the uptake of the technology and services. There are numerous applications and opportunities in this area:

- Production line efficiency by making sure that the correct tools and equipment are available and serviceable at all times through the manufacturing process – precise positioning technologies, such as UWB, are widely used in the automotive and aeronautical industries;
- Keeping track of valuable critical assets, such as medical equipment in hospitals, to make sure that facilities are available when needed and to minimise lost time searching for equipment;

- Operational efficiency in underground mines by monitoring both miners and machines, thereby ensuring that the correct resources are deployed in the right places and that downtime is minimised by reducing dead time and minimising the risk of collisions;
- Construction is reliant on a complex material supply chain and coordinating it with the availability of skilled workers for each task in the project plan – tracking materials, equipment and people through the whole process can help to improve efficiency;
- Livestock management, such as dairy farming, can benefit greatly with improved herd management – optimising feed regimes, reducing 'downtime' through illness by improved herd welfare, more efficient fertility detection and more efficient management of slurry and general farm operations – all of which can be improved through the use of positioning, location and behavioural monitoring technologies;
- Freight logistics, particularly at modal interchange points (loading and unloading) can be streamlined by technology that helps to manage the process to reduce waiting times and risk of sending a load to the wrong destination;
- Ground vehicle management around airports and container ports, as well as baggage handling and passenger management can all benefit from positioning and location information;
- Retail space optimisation based on shopper behavioural patterns.

9.3.8 Sport

Sport has always been characterised by timing, speed and position. Technologies that help with tracking, positioning, timing and behavioural measuring are widely adopted for use in sport:

- Spectators, supporters and sporting fans have a real desire for better and more detailed information about the sport, not just times and winners, but also statistics of all kinds such as peak speed, distance covered, pulse rate, g-force etc.;
- Performance information about the athletes, vehicles or equipment can be used to improve performance and competitiveness through training regimes;

- Information about speed, position and direction of participants, balls and other equipment can be used by officials to help with decisions regarding outcome and/or rule infringement;
- Speed, position and orientation information can be used by broadcasters to improve coverage of the event by assisting with camera selection and control.

Positioning and location technologies are already widely used in many sports such as motor racing, horse racing, ocean sailing and even certain track events like marathon running.

10 A brief look at the future

Whilst it is impossible to predict the future there are a number of trends at the cutting edge of location and positioning that tell us what the topical issues of today are and hint at the direction things may go in the future.

Firstly there is the explosion in new GNSSs coming online. Combining measurements from them will lead to improved performance as well as robustness and ubiquity. However, we have shown that GNSS is vulnerable to attack and there is also some debate about whether so many systems will begin to interfere with one another. So whilst GNSS technology continues to advance there is also a significant amount of work going into relative positioning systems for special applications. These systems combine the best of radio signals and other sensor measurements in order to extract meaningful information about the relative position and location context of people and objects. In high-end general applications these systems will be combined with GNSS in order to provide seamless coverage indoors and out.

Of course on the technology side we must not forget about visual and imaging systems. These mimic the way we as humans perceive our environment better than any other, and with ongoing rapid advancement of image processing technology they promise to be a key component of future positioning and localisation systems.

However, positioning and localisation is not only about computing an x,y,z position – the 'information' arising from this knowledge is arguably the most important aspect. There remains a great deal of work to be done in this area, and we are still a long way from being able to adequately describe relative positions between objects or the context of an indoor location.

10.1 GNSS advances

10.1.1 More constellations

At the time of writing GPS and Glonass are fully operational systems.

The Chinese system COMPASS was declared partially operational for test use in the Asia region in December 2011. Additional satellites are being launched on a regular basis and the system is scheduled to provide global operational capability by 2020.

Galileo, the European GNSS, has been plagued by delays but satellites are now being launched, with two in orbit at the end of 2011, on a timeline that should lead to initial operational capability by 2015 (18 satellites). It is anticipated that full operational capability will be reached by about 2020.

QZSS (Quasi-Zenith Satellite System) being implemented by Japan is designed primarily to complement and reinforce GPS (and Compass and Galileo in due course), by providing a constellation of up to seven additional satellites high in the sky above Japan. The primary role is to provide augmentation data that allows for quicker and more reliable acquisition of navigation signals and which provides correction and integrity data to improve positioning accuracy. However, the system also provides limited navigation capability when used alone. The first satellite is operational and budgetary approval to continue with system deployment has been secured.

IRNSS (Indian Regional Navigation Satellite System) is an independent GNSS providing services to the Indian Region. It comprises three geostationary satellites and four satellites in GEO synchronous orbit with longitude crossings at 55° and 111° East, making all the satellites visible from Indian ground control stations. Dual frequency operation gives an estimated positioning accuracy of better than 20 metres in the Indian subcontinent region. The first satellites are planned for launch in 2013 or 2014.

There is some concern that multiple GNSSs could lead to mutual interference leading to poorer performance rather than the more widely heralded benefits of more satellites and more signals leading to improved accuracy and better coverage of obscured areas. Recognising this there are significant international efforts being devoted to encourage interoperability and compatibility between systems. Compatible systems do not interfere with each other but they operate largely independently.

Interoperability is a much stronger condition and allows a receiver to combine signals from satellites belonging to different systems. To achieve interoperability the different systems should have compatible and consistent system time and geodetic coordinate systems. This is being achieved through time source traceability to UTC (Coordinated Universal Time) and the use of coordinate systems aligned to the ITRS (International Terrestrial Reference System).

Even stronger than interoperability is interchangeability, which would allow the receiver to compute a position (and time offset) using any four satellites from any of the available constellations. Doing this requires the different systems to have synchronised clocks, and common carrier frequency. GPS, Compass, Galileo and QZSS share two common carrier frequencies and Glonass is considering this possibility.

10.1.2 New types of signals

Next generation (modernised) GPS and the new systems (Galileo and Compass) are introducing a new frequency band (L5) as well as new signal modulations in the existing bands. These include more advanced BOC (binary offset carrier) modulation, longer code lengths with repeat intervals of 50 ms and 1500 ms, and carrier-only signals without data modulation.

The main advantages these various improvements bring are easier (and quicker) signal acquisition using weaker signals. The ability to integrate coherently for longer periods of time leads to improved accuracy, subject to dynamic motion behaviour of the tracked object.

10.1.3 Better receivers

Improvement in receiver technology is one of the main factors behind significantly improved performance over the past decade. Multichannel capability means that receivers can track all the satellites in view simultaneously and they no longer need to multiplex between satellites. Increasing computational capability in electronic components and reducing power requirements mean that more processing can be done using less power. Although the true software GPS receiver is not yet commercially viable, it is only a matter of time before it will be.

It is highly likely that significant further improvements in receiver design will be achieved over the next decade. This may be achieved through better signal tracking using more sophisticated tools, maximum likelihood algorithms, Bayesian filtering, and other methods. It may be achieved through smarter digitising radio front ends, or it may be achieved through better and tighter integration with sensor measurements – such as from inertial MEMS sensors. However, the proliferation of new signals and satellite systems using new modulation and coding schemes will demand more advanced receivers just to acquire and use the signals, let alone to achieve additional benefits of shorter time-to-first-fix, improved accuracy or lower power.

10.1.4 Augmentation services

Augmentation is a term used to describe methods of improving the performance of a satellite navigation system: not just accuracy, but also reliability and availability. Augmentation systems provide additional information about the performance and behaviour of the GNSS obtained externally into the calculation process.

Perhaps most important is the integrity monitoring and fault detection capability that these systems provide.

Generally augmentation systems can be categorised into satellite-based and ground-based systems.

10.1.4.1 Satellite-based augmentation systems (SBAS)

In a satellite-based augmentation system special receivers on the ground receive and measure the satellite signals. Since the receivers are installed at known positions they are able to analyse the received signals to determine system errors. These include both signal propagation errors and satellite errors. Measurement corrections are relayed via geostationary satellites which broadcast the corrections to any augmentation-enabled receivers on the ground. These receivers receive the broadcast corrections and apply them to their received measurements thereby improving the receiver performance.

There are several satellite augmentation systems in operation or being proposed, including:

- Wide Area Augmentation System (WAAS) operated by the FAA and covering the North American region;
- European Geostationary Navigation Overlay Service (EGNOS) operated by the European Space Agency and covering the European region;
- Multi-functional Satellite Augmentation System (MSAS) operated by Japan's Ministry of Land, Infrastructure and Transport;
- Quasi-Zenith Satellite System (QZSS), proposed by Japan;
- GAGAN (GPS and Geo-Augmented Navigation) system, proposed by India;
- Russian System for Differential Correction and Monitoring (SDCM), proposed by Russia;
- Satellite Navigation Augmentation System (SNAS), proposed by China.

10.1.4.2 Ground-based augmentation systems

Augmentation signals may also be transmitted by ground stations rather than by satellites. This is commonly used in surveying systems that use differential GPS and RTK (real-time kinematic) techniques, although the corrections are usually generated and used privately within a closed network containing its own reference receiver.

There are also a number of public sources for correction data, delivered by ground systems, the following being just a few of those in operation:

- A European DGPS network has been developed largely by Finnish and Swedish maritime associations to improve safety in the archipelago in the region.

- In the UK a number of Trinity House DGNSS stations broadcasting in the 300 kHz band are operated to aid maritime navigation around the Isles.
- A network of European Differential Beacon Transmitters is being set up around most of the European coastline as an extension to the UK system and also operating in the 300 kHz band.
- The US Nationwide Differential GPS System (NDGPS) is a ground-based augmentation system transmitting correction in the LF band and modelled on the European system.
- Australia runs several DGPS systems using the long-wave band as well as commercial FM radio to transmit corrections.

10.1.4.3 Assisted GPS (A-GPS)

Assisted GPS is a technique used primarily to improve the time-to-first-fix of a GPS receiver. It is used extensively with mobile cellular phones. The principle is that a network-based server is used to supply information that allows much quicker acquisition of the satellite signals by the receiver. This information may comprise ephemeris and almanac for the satellites and precise time. This reduces the signal search space in terms of codes, time and frequency and therefore allows the receiver to acquire the satellite signal much quicker.

A-GPS systems may also operate in a network mode in which the receiver supplies measurements, or a snapshot of the received signal, to a central processing device which computes the position of the receiver.

Signalling between the mobile device and assistance server may be done using either control plane protocols or user plane protocols:

- 3GPP and 3GPP2 have standardised control plane (signalling) protocols for the exchange of assistance data between a network provided assistance server and the mobile device.
- OMA (Open Mobile Alliance) has defined the SUPL (secure user plane location) protocol for exchanging assistance data with the mobile terminal. User plane protocols make use of a standard packet switched connection between the mobile device and server and are therefore independent of network operator and infrastructure, provided that the user terminal supports data services.

10.1.5 Precise point positioning (PPP)

PPP is based on the use of both code and carrier phase measurements in either a single or dual frequency receiver which when combined with accurate ephemeris and clock data for the satellites, allows the receiver position to be determined to high accuracy – tens of centimetres. There are a number of online sources from which correction data can be obtained, typically offering 10 cm accuracy position corrections and 1.5 ns accuracy clock corrections.

Compared to RTK systems, PPP operates using a single receiver, with no need for a reference base station. A description of the principles of PPP can be found in Zumberge [51]. The PPP receiver needs to go through a fairly lengthy calibration and convergence period of, typically, 20 minutes before it is ready for operation. It also needs to obtain, typically via a satellite augmentation service, or an internet service provider, ephemeris and clock correction data for the satellites.

PPP has so far been mainly used for offline processing of captured satellite measurements, and a number of online services to which measurements can be uploaded and the results delivered back to the user by email, or download, exist for this purpose. However, real-time implementations are emerging and PPP has already been adopted for use in agriculture amongst other applications.

It is likely that PPP algorithms used in conjunction with high accuracy augmentation services will lead to significant GNSS performance improvements over the next decade. The IGS (International GNSS Service) is a voluntary collaboration of more than 200 contributing organisations in more than 80 countries which provides precise GPS orbit products to the scientific community – to millimetre precision and 0.03 ns clock estimates for some products. The IGS Products Guide [52] contains more information.

10.1.6 Interference and jamming mitigation

GPS receivers are susceptible to interference whether intentional or not. As the use of computing and communications becomes more widespread the likelihood of suffering from interference increases and with the widespread use of the GPS in numerous applications, the impact of poor performance is much more serious. The problems are not unique to GPS and equally affect all other existing and proposed GNSS systems.

One of the areas in which significant future advances will take place is the tolerance of receivers to interferers. These roughly divide into:

- avoidance techniques, antennas, improved GPS installation, and use of alternative positioning technologies;
- law enforcement, localisation of interference sources and prosecution of offenders through the legal process;
- advanced receiver and signal processing techniques.

10.1.6.1 Avoidance techniques

The FAA has already taken a number of steps to ensure that the GPS receivers used in aircraft landing systems are hardened against GPS interference and jamming threats. These mostly fall into the category of avoidance techniques:

- Use of special antennas with directional characteristics that help avoid interference – such as very good attenuation of signals from ground-based sources, or with steerable beams and/or nulls. An example of a commercial product offering this capability is the Novatel GAJT.
- Better placement of GPS antennas. At airports and cellular base station sites this could mean placing them higher up, or further from public areas and roads.
- Use of secondary positioning technologies such as eLoran or, for vehicles, inertial navigation sensors. For cellular base stations and in timing applications this may involve the use of cabled (copper or fibre optic) time distribution and more stable free-running clocks.

10.1.6.2 Law enforcement

Considerable effort has been put into techniques to identify sources of interference and prosecute offenders in the courts – for example the GAARDIAN project in the UK and JLOC in the USA.

So far these systems and techniques are relatively immature and it has proven very difficult to identify and track down sources of interference.

However, with continuing development in this area it is likely that methods of detection will be greatly improved leading to increased number of prosecutions of offenders.

10.1.6.3 Advanced receiver design

Advanced receiver designs that are better able to mitigate signal interference are only just becoming available. This is potentially the area in which the greatest future improvements will be achieved, but it has been held back by restrictions on being able to test receiver performance with real-world interfering signals.

Better quality A/D converters can provide improved immunity against CW interferers (these are the most common type of unintentional interference, often arising as an unwanted harmonic spur from an oscillator in an electronic device).

Integration with inertial sensors can allow smarter more adaptive and narrower tracking loops to be used, thereby maintaining dynamic receiver performance and giving an estimated 10 to 20 dB improvement in interference rejection.

Adaptive and steerable antennas giving spatial signal selectivity and adaptive temporal filters with tunable rejection notches can significantly improve rejection of CW interferers. Whilst a number of techniques have been demonstrated, and some even adopted commercially in high-end receivers, there is considerable scope for advanced signal processing algorithms and techniques.

Each of these steps is progressively more complex, and as such they carry cost and power penalties, but for many mission-critical applications the cost is small compared to the risk and it is expected that many advanced techniques will emerge in the future.

10.2 Relative positioning systems

10.2.1 Cooperative use of measurements of neighbouring devices

The fundamental principle behind traditional radiolocation systems starts with the principle that a number of devices – either transmitters or receivers – are installed at known locations. This is how GPS works, the fact that the satellites are moving is merely a point of detail – their positions are known at all times. The device to be located either measures the signals from devices at known positions, or the known devices measure the signals transmitted by the devices at unknown positions. In this way the position of a mobile device can be determined.

This basic system requires the clocks in the devices at known positions to be accurately synchronised. Clock error translates into positional uncertainty based on the speed of light which represents about 1 foot of error (0.3 metres) for every 1 ns of clock timing error. Next generation systems are able to work with free-running unsynchronised clocks by using local measurement units (LMUs) installed at known locations. Using these LMUs the clocks of the fixed devices can be measured and characterised and the bias subtracted from measurements of mobile devices. This is the principle used in mobile cellular systems of E-OTD and OTDOA.

The next evolutionary step was the development of a system by Cambridge Positioning Systems, which they called 'Matrix' in which they were able to eliminate the need for LMUs by collectively using measurements from multiple mobile terminals in order to compute the clock offsets of the cellular base stations by exploiting the time-space diversity of these measurements.

Another approach used for positioning unknown devices in a local network is to use ranges measured by exchanging two-way signals between pairs of transceivers. These systems are generally based on range measurements between a mobile device having unknown position and static devices having known positions. These systems have been shown to work well in small-scale networks, but the need to make multiple point–point range measurements seriously limits their ability to scale.

An evolution of the system based on point–point range measurements in which all the devices are mobile has been demonstrated. Such a system enables a group of identical mobile devices to determine their positions relative to one another. The main advantage such systems offer over conventional approaches is the 'flattening' of the architecture by eliminating the distinction between fixed devices (sometimes called access points or anchors) and mobiles, thereby making for easier installation and cheaper systems.

Research and development into relative positioning systems is an active ongoing area. By replacing point-to-point range measurements with time-of-arrival measurements made and shared between a network of similar transceiver devices it is possible for them to determine their relative positions, and the use of broadcast signals for which times of arrival are measured by all (or many) neighbours within range allows such systems to scale to large networks comprising many devices.

A further extension to the use of neighbour measurements for relative positioning is the use of shared and collective GNSS signal measurements. In the simplest form local relative positioning could allow the extension of location to devices without GNSS signal coverage based on the positions of neighbours able to compute GNSS positions. Taking this approach further, devices with GNSS coverage (including partial coverage) can make use of GNSS satellite measurements made by neighbours to help locate themselves given knowledge of the relative positions of the distributed receivers.

It is likely that many applications in the future will benefit from and make use of collective, peer and neighbour measurements within a relative context as a means to extending the ubiquity and reliability of locating techniques.

10.2.2 Using signals of opportunity

Another emerging technique for locating devices is the use of signals of opportunity. This is an extension of fingerprinting techniques in which a device compiles a 'map' of observed signals against which it compares future measurements in order to determine that it has returned to a particular previously visited place.

This is an example of a learning system in which its sensors measure and monitor the radio environment to construct a fingerprint of the location. This technique is already widely used for Wi-Fi signals but with more flexible radios emerging it could be extended to use a wide range of radio signals from many bands including mobile cellular signals, radio and TV transmissions and others.

10.2.3 Advanced inertial sensors

The use of inertial sensors to provide navigation and positioning assistance is already becoming widespread with the emergence of good quality low-cost MEMs devices. However, there is still considerable scope for improvements in both sensor performance and the algorithms to exploit measurements from the sensors, and this will continue to be an important area of growth.

10.3 Visual and optical positioning systems

Optical techniques are already widely used for ball tracking in sport and photogrammetry as a technique for combining stereoscopic images to extract height information has been used for decades. However, the game changer happening now is the proliferation of low-cost cameras and imaging devices fitted to a large proportion of consumer devices (phones, tablet computers etc.).

When combined with increased processing power available and low-cost memory and storage these imaging devices (cameras) will open up many new and innovative locating and positioning capabilities and services. In the same way that we know where we are by recognising places and landmarks, imaging devices will eventually be able to do the same, and also a lot more.

10.4 Privacy issues

Privacy is one of those subjects which divides opinions like no other. On the one hand opponents see widespread availability of location information as an invasion of privacy. On the other hand others see knowledge of location as being empowering. Only one thing is certain: privacy-related issues will continue to dog the localisation and positioning industry for many years.

The subject of privacy is far too extensive to cover in detail here, and the reader is referred to some of the good reference works on the subject [53] [54] [55] [56]. It is a key issue with which future location-based services and applications will have to deal.

Location information describes the where and when of someone or something. It has been said that it is not about where you are but about where you've been. Current laws are relatively open and vague when it comes to location information. The Information Commissioner's Office in the UK has called for a rethink on location privacy to avert future consumer backlash. In California considerable lobbying against the proposed California Location Privacy Bill (SB 1434) has begun on the basis that it gives too much freedom to operators and enforcement authorities to obtain and use location information. Google and Apple's widespread collection and use of location data have led to adverse publicity for both over the past months and years. However, the availability of this information has also been an enabler for many new and exciting applications and services.

One of the major organisations lobbying for stronger location privacy laws is the Electronic Frontier Foundation [57].

Privacy is an issue for the location industry as a whole and it will not be solved quickly. Quite likely attitudes to privacy and location information are generational and in time we will see shifts in the way it is collected, used and shared.

10.5 The information of positioning

This book started by describing how location and position have traditionally been defined in a navigational sense, typically as a latitude, longitude and height, or similar global coordinate system. However, in the context of modern positioning applications and needs which are geared to the way we as humans think and behave location and positioning are more about context and relationships between us and our environment. In order for location and positioning applications to become ubiquitous it will be essential to find natural ways for describing positioning information, especially for local area and indoor applications.

Extending location information to provide more than x,y,z will be essential. Orientation (direction) will be fundamental as will ways to recognise the temporal (time-based) characteristics of location, without which behaviours cannot be fully described.

References

[1] N. Samama, *Global Positioning Technologies and Performance*, Hoboken, New Jersey: Wiley, 2008.

[2] T. Gooley, *The Natural Navigator*, 1st edn, UK: Virgin Books, 2010.

[3] T. Gooley, *The Natural Navigator: A Watchful Explorer's Guide to a Nearly Forgotten Skill*, 1st edn, USA: The Experiment, 2011.

[4] P. F. Mottelay, *Bibliographical History of Electricity and Magnetism*, London: Charles Griffin And Company Limited, 1922, http://www.archive.org/details/bibliographicalh033138mbp

[5] D. Gubbons and E. Herrero-Bervera (eds.), *Encyclopedia of Geomagnetism and Paleomagnetism*, The Netherlands: Springer, 2007.

[6] N. Bowditch, *The New American Practical Navigator*, 20th edn, New York: E. & G. W. Blunt, 1851.

[7] N. Bowditch, *The American Practical Navigator: an Epitome of Navigation*, originally by Nathaniel Bowditch in 1802, revised and updated, Bethesda, Maryland: National Imagery and Mapping Agency, 2002.

[8] J. Karl, *Celestial Navigation in the GPS Age*, Paradise Cay Publications, 2009.

[9] D. Sobel, *Longitude*, London: Fourth Estate, 1996.

[10] Ordnance Survey, *A Guide to Coordinate Systems in Great Britain*, D00659 v2.1 Dec 2010.

[11] C. F. Gauss (Author), A. Hiltebeitel (Translator), J. Morehead (Translator), *General Investigations of Curved Surfaces*, Wexford College Press, 2007.

[12] J. Snyder, *Flattening the Earth: Two Thousand Years of Map Projections*, University of Chicago Press, 1998.

[13] *Military Map Reading*, http://earth-info.nga.mil/GandG/coordsys/mmr201.pdf

[14] J. B. Kuipers, *Quarternions and Rotation Sequences*, Princeton University Press, 1998.

[15] *Navstar GPS Space Segment/Navigation User Interfaces*, IS-GPS-200, Revision D, GPS Navstar Joint Program Office, El Segundo, CA, 7 December 2004.

[16] R. Gold, 'Optimal binary sequences for spread spectrum multiplexing'. *IEEE Transactions on Information Theory*, vol. 13, no. 4, pp. 619–621, October 1967.

[17] G. X. Gao, L. Heng, D. De Lorenzo *et al.*, 'Modernization milestone: observing the first GPS satellite with an L5 payload'. *InsideGNSS*, May/June 2009.

[18] E. D. Kaplan and C. J. Hegarty, *Understanding GPS: Principles and Applications*, 2nd edn, Norwood: Artech House Mobile Communications Series, 2006.

[19] *BeiDou Navigation Satellite System: Signal In Space: Interface Control Document (Test Version)*, China Satellite Navigation Office, December 2011. Available from http://www.beidou.gov.cn/attach/2011/12/27/201112273f3be6124f7d4c7bac428a36cc1d1363.pdf

[20] *Global Navigation Satellite Systems: Report of a Joint Workshop of the National Academy of Engineering and the Chinese Academy of Engineering*, National Academy of Engineering, 2012, http://www.nap.edu/catalog.php?record_id=13292

[21] I. Amundson, J. Sallai, X. Koutsoukos and A. Ledeczi, 'Radio interferometric angle of arrival estimation'. *EWSN 2010*, pp. 1–16. Springer. Retrieved from http://www.isis.vanderbilt.edu/node/4139

[22] W. Li, W. Yao and P. J. Duffett-Smith, 'Comparative study of joint TOA/DOA estimation techniques for mobile positioning applications'. *6th IEEE Consumer Communications and Networking Conference*, pp. 1–5, 2009.

[23] J. Sallai, G. Balogh, M. Maróti, Á. Lédeczi and B. Kusý, 'Acoustic ranging in resource-constrained sensor networks'. *International Conference on Wireless Networks*, pp. 467–472, 2004.

[24] W. H. Press, S. A. Teukolsky, W. T. Vetterling and B. P. Flannery (eds.), *Numerical Recipes: The Art of Scientific Computing*, Cambridge University Press, 2007.

[25] J. S. Lim and A. V. Oppenheim (eds.), *Advanced Topics in Signal Processing*, Englewood Cliffs, New Jersey: Prentice-Hall, 1988.

[26] D. Humphrey and M. Hedley, 'Super-resolution time of arrival for indoor localization'. *IEEE International Conference on Communications, 2008. ICC '08*, pp. 3286–3290, 19–23 May 2008.

[27] R. W. Rowe, P. J. Duffett-Smith, M. R. Jarvis, N. G. Graube and C. Silicon, 'Enhanced GPS: the tight integration of received cellular timing signals and GNSS receivers for ubiquitous positioning'. *ION PLANS 08*, pp. 828–845, 2008.

[28] P. Duffett-Smith and M. Macnaughtan, 'Precise UE positioning in UMTS using cumulative virtual blanking'. *3G Mobile Communication Technologies 2002 Third International Conference*, conf. publ. no. 489, vol. 2002, pp. 355–359, IET, 2002.

[29] C. Drane, P. J. Duffett-Smith, S. Hern and J. Brice, 'Mobile positioning using E-OTD without LMUs'. *Proceedings of AeroSense 2003, the International Society for Optical Engineering's (SPIE's) 7th Annual International Symposium on Aerospace/Defense Sensing, Simulation, and Controls*, Orlando, Florida, 21–25 April 2003.

[30] Z. Sahinoglu, S. Gezici and I. Güvenc, *Ultra-wideband Positioning Systems: Theoretical Limits, Ranging Algorithms, and Protocols*, Cambridge University Press, 2008.

[31] D. Titterton and J. Weston, *Strapdown Inertial Navigation Technology*, 2nd edn, London: The Institution of Electrical Engineers, 2004.

[32] M. S. Grewal, L. R. Weill and A. P. Andrews, *Global Positioning Systems, Inertial Navigation, and Integration*, 2nd edn, Hoboken, New Jersey: Wiley, 2007.

[33] P. Harrop and G. Holland, *Real Time Locating Systems (RTLS) 2008–2018*, IDTechEx Ltd.

[34] M. A. Quddus, W. Y. Ochieng and R. B. Noland, 'Current map-matching algorithms for transport applications: state-of-the art and future research directions'. *Transportation Research Part C: Emerging Technologies*, vol. 15, no. 5, pp. 312–328, 2007.

[35] R. Sim, P. Elinas, M. Griffin and J. J. Little, *Vision-based SLAM using the Rao-Blackwellized Particle Filter*, Laboratory for Computational Intelligence, University of British Columbia, Vancouver, BC, V6T 1Z4, Canada, 2010.

[36] G. Retscher, E. Moser, D. Vredeveld, D. Heberling and J. Pamp, 'Performance and accuracy test of a WiFi indoor positioning system'. *Journal of Applied Geodesy*, vol. 1, no. 2, pp. 103–110, 2007.

[37] C. E. Rasmussen and C. K. I. Williams, *Gaussian Processes for Machine Learning*, MIT Press, 2006.

[38] B. Ferris, D. Fox and N. Lawrence, 'WiFi-SLAM using Gaussian process latent variable models'. *20th International Joint Conference on Artificial Intelligence*, January 2007, pp. 2480–2485.

[39] S. Tennina, M. Di Renzo, F. Santucci and F. Graziosi, 'On the distribution of positioning errors in wireless sensor networks: a simulative comparison of optimization algorithms'. *Wireless Communications and Networking Conference 2008*, IEEE, pp. 2075–2080, 2008.

[40] L. Weill, 'C/A code pseudoranging accuracy – how good can it get?' *Proceedings of ION GPS-94, the 7th International Technical Meeting of the Satellite Division of the Institute of Navigation*, pp. 133–141, 1994.

[41] Tao Jia and R. M. Buehrer, 'A new Cramér–Rao lower bound for TOA-based localization'. National Science Foundation Grant 0515019, IEEE, 2008.

[42] S. Reece and D. Nicholson, 'Tighter alternatives to the Cramér–Rao lower bound for discrete-time filtering'. *8th International Conference on Information Fusion*, IEEE, 25–28 July 2005.

[43] B. Ristic, S. Arulampalam and N. Gordon, *Beyond the Kalman Filter: Particle Filters for Tracking Applications*, Norwood: Artech House.

[44] S. Pullen and G. Xingxin Gao, 'GNSS jamming in the name of privacy – potential threat to GPS aviation'. *InsideGNSS*, March/April 2012, pp. 34–43.

[45] R. H. Mitch, R. C. Dougherty, M. L. Psiaki *et al.*, 'Innovation: know your enemy'. *GPS World*, 1 January 2012.

[46] A. Grant, *GPS Jamming Trial*, Research & Radionavigation, General Lighthouse Authorities, UK & Ireland, RPT-26-AJG-08, 23/09/2008.

[47] O. Pozzobon, 'Keeping the spoofs out: signal authentication services for future GNSS'. *InsideGNSS*, May/June 2011, pp. 48–55.

[48] *NMEA 0183, The Standard for Interfacing Marine Electronics*, National Marine Electronics Association, Version 4.10, November 2008.

[49] *Road Pricing Demonstrations Project*, Department for Transport (DfT), Report 23 June 2011, http://www.dft.gov.uk/publications/road-pricing-demonstrations-project

[50] 'Urban transport pricing in Europe', http://www.transport-pricing.net/

[51] J. F. Zumberge, M. B. Heflin, D. C. Jefferson and M. M. Watkins, 'Precise point positioning for the efficient and robust analysis of GPS data from large networks'. *Journal of Geophysical Research*, vol. 102, no. B3, pp. 5005–5017, 1997.

[52] J. Kouba, *A Guide to Using International GNSS Service (IGS) Products*, Geodetic Survey Division, Natural Resources Canada, May 2009, http://igscb.jpl.nasa.gov/igscb/resource/pubs/UsingIGSProductsVer 21.pdf

[53] J. Waldo, H. S. Lin and L. I. Millett (eds.), *Engaging Privacy and Information Technology in a Digital Age*, Committee on Privacy in the Information Age, Computer Science and Telecommunications Board, Division on Engineering and Physical Sciences, National Research Council of the National Academies, The National Academies Press, Washington, DC, 2007.

[54] P. E. Agre and M. Rotenberg, *Technology and Privacy: The New Landscape*, MIT Press, 1998.

[55] D. H. Holtzman, *Privacy Lost: How Technology Is Endangering Your Privacy*, John Wiley & Sons, 2006.

[56] J. W. DeCew, *In Pursuit of Privacy: Law, Ethics, and the Rise of Technology*, Cornell University Press, 1997.

[57] A. J. Blumberg and P. Eckersley, *On Locational Privacy, and How to Avoid Losing it Forever*, Electronic Frontier Foundation, August 2009, https://www.eff.org/wp/locational-privacy/

Index